Sustainability and Environmental Impact of Renewable
Energy Sources

ISSUES IN ENVIRONMENTAL SCIENCE AND TECHNOLOGY

ISSUES IN ENVIRONMENTAL SCIENCE
AND TECHNOLOGY

EDITORS: R. E. HESTER AND R. M. HARRISON

19

Sustainability and Environmental Impact of Renewable Energy Sources

RS•C

advancing the chemical sciences

ISBN 0-85404-290-3
ISSN 1350-7583

A catalogue record for this book is available from the British Library

Published by The Royal Society of Chemistry, Thomas Graham House,
Science Park, Milton Road, Cambridge CB4 0WF, UK

Registered Charity Number 207890

For further information see our web site at www.rsc.org

Typeset in Great Britain by Vision Typesetting, Manchester
Printed and bound by Bookcraft Ltd, UK

Preface

The 2002 World Summit on Sustainable Development in Johannesburg focused attention of the world's media on environmental issues, important among which was the development of renewable energy (RE) sources. These are widely seen as a means of combating climate change induced by rising levels of CO_2 in the atmosphere resulting from the combustion of fossil fuels. The rate of energy consumption is predicted to triple over the next 50 years or so, emphasizing the need to develop novel sources with lower environmental impact.

Given that the current overwhelming dependence on coal, oil and gas for power generation, heating and transport is likely to continue for some considerable time into the future, there is need for increasing attention to be given to improving efficiency and the development of methods for carbon sequestration. The role of nuclear power is also under review. It is particularly timely then, to examine the prospects for the so-called renewables, such as wind, wave and tidal power, thermal and photovoltaic solar, hydro, biomass and biogas. We have commissioned a group of leading experts to contribute authoritative reviews of the various aspects of the subject in a critical but balanced way.

The first article is by Bernard Bulkin, Chief Scientist at BP plc. Taking as its starting point an analysis of the changes resulting from the last 'energy crisis' of 1973, this reviews 'The Future of Today's Energy Sources' and so provides a basis for evaluation of the emerging alternatives, including renewables. The future supply and demand for coal, oil and gas in the various sectors in which they are used, *viz.* transport, power generation and industrial/domestic/agricultural applications, are examined. This review concludes that shortage of supply will not be a driver for change in the near future and that increases in the efficiency of use and the development of effective methods for carbon sequestration will prolong the role of fossil fuels.

Further evaluation of the prospects for energy conservation and carbon sequestration is included in the second article. Together with a critical assessment of the role of nuclear power, this sets the scene for a shift in focus to the new renewable energy technologies. In his article on 'Sustainable Energy: Choices, Problems and Opportunities', David Elliott of the Open University's Energy and Environment Research Unit examines the economic, social and environmental implications of the development of sustainable energy systems and the strategic technological choices that lie ahead. The article is rich in detail on comparative costs and risks.

The integration of renewable energy generation into the electricity supply system without compromising safety or reliability is key to the commercial exploitation of RE. The technological considerations associated with integration are analysed in the next article by David Infield and Paul Rowley of Loughborough University's Department of Electronic and Electrical Engineering. The variability of many of the RE sources (*e.g.* wind) presents problems for integration, but these can largely be overcome by aggregation of supply (*e.g.* wind farms) and demand (many consumers). The technical issues associated with network integration of wind power, photovoltaics, water power and biomass are outlined here and ways of dealing with the intrinsic problems of stand-alone systems are described.

The fourth article, written by Adrian Leoning of CPL Envirogas Ltd, is concerned with the potential for power generation from freely available organic material. This encompasses landfill gas ('energy from waste'), biomass and waste bio-oils, and includes evaluation of both technological and economic considerations. Then follows an article by Fiona Mullins, who is a Senior Associate of Environmental Resources Management in Oxford. This addresses emissions trading schemes and their consequences on reduction/redistribution of environmental burdens and impacts. It concludes that emissions trading is emerging as a 'licence to pollute' in the early stages, but is likely to be an effective tool for reducing greenhouse gases in the longer term. An article by the current UK government Minister for Energy and Construction, Brian Wilson, then outlines the government's renewable energy policy, providing insights into the workings of the Non-Fossil Fuel Obligation (NFFO) and other related policy tools.

The final article, by Andrew Stirling of the Science and Technology Policy Research Unit (SPRU) at the University of Sussex, offers a critical analysis of the methods currently in use for the appraisal of sustainability of the various energy options. Conventional comparative risk assessment and environmental cost–benefit analyses of renewable and other electricity supply technologies are shown to be flawed. The relationship between these established 'science-based' techniques and the newly emerging 'precautionary' approaches to sustainability appraisal is discussed and some practical ways forward are presented.

Overall, we believe that this collection of articles provides a timely and authoritative examination of many of the most important issues and concerns relating to renewable energy sources. The extent to which these offer solutions to problems of climate change and sustainability is under intense scrutiny now and the insights provided by these review articles make them essential reading for all concerned with the environment. The volume is commended particularly to those engaged in the energy industries, both traditional and developing, to policy makers, consultants and engineers working in this field, and to both teachers and students in higher education.

Ronald E. Hester
Roy M. Harrison

Contents

Issues in Environmental Science and Technology, No. 19
Sustainability and Environmental Impact of Renewable Energy Sources
© The Royal Society of Chemistry, 2003

Contents

Contents

Contents

Editors

Ronald E. Hester, BSc, DSc(London), PhD(Cornell), FRSC, CChem

Ronald E. Hester is now Emeritus Professor of Chemistry in the University of York. He was for short periods a research fellow in Cambridge and an assistant professor at Cornell before being appointed to a lectureship in chemistry in York in 1965. He was a full professor in York from 1983 to 2001. His more than 300 publications are mainly in the area of vibrational spectroscopy, latterly focusing on time-resolved studies of photoreaction intermediates and on biomolecular systems in solution. He is active in environmental chemistry and is a founder member and former chairman of the Environment Group of the Royal Society of Chemistry and editor of 'Industry and the Environment in Perspective' (RSC, 1983) and 'Understanding Our Environment' (RSC, 1986). As a member of the Council of the UK Science and Engineering Research Council and several of its sub-committees, panels and boards, he has been heavily involved in national science policy and administration. He was, from 1991 to 1993, a member of the UK Department of the Environment Advisory Committee on Hazardous Substances and from 1995 to 2000 was a member of the Publications and Information Board of the Royal Society of Chemistry.

Roy M. Harrison, BSc, PhD, DSc(Birmingham), FRSC, CChem, FRMetS, Hon MFPHM, Hon FFOM

Roy M. Harrison is Queen Elizabeth II Birmingham Centenary Professor of Environmental Health in the University of Birmingham. He was previously Lecturer in Environmental Sciences at the University of Lancaster and Reader and Director of the Institute of Aerosol Science at the University of Essex. His more than 300 publications are mainly in the field of environmental chemistry, although his current work includes studies of human health impacts of atmospheric pollutants as well as research into the chemistry of pollution phenomena. He is a past Chairman of the Environment Group of the Royal Society of Chemistry for whom he has edited 'Pollution: Causes, Effects and Control' (RSC, 1983; Fourth Edition, 2001) and 'Understanding our Environment: An Introduction to Environmental Chemistry and Pollution' (RSC, Third Edition, 1999). He has a close interest in scientific and policy aspects of air pollution, having been Chairman of the Department of Environment Quality of Urban Air Review Group and the DETR Atmospheric Particles Expert Group as well as a member of the DEFRA Expert Panel on Air Quality Standards. He is currently a member of the DEFRA Air Quality Expert Group, the DEFRA Advisory Committee on Hazardous Substances and the Department of Health Committee on the Medical Effects of Air Pollutants.

Contributors

B.J. Bulkin, *BP plc, Brittanic House, 1 St James's Square, London SW1Y 4PD, UK*

D.A. Elliott, *Energy and Environment Research Unit, The Open University, Milton Keynes MK7 6AA, UK*

D.G. Infield, *Department of Electronic and Electrical Engineering, Loughborough University, Loughborough LE11 3TU, UK*

A. Loening, *CPL Envirogas Ltd, 20–22 Queen Street, London W1X 7PJ, UK*

F. Mullins, *Environmental Resources Management, Fernwell House, 76 Harpes Road, Oxford OX2 7QL, UK*

P.N. Rowley, *Department of Electronic and Electrical Engineering, Loughborough University, Loughborough LE11 3TU, UK*

A.C. Stirling, *Science and Technology Policy Research Unit, Mantell Building, University of Sussex, Brighton BN1 9RF, UK*

B. Wilson, *Minister for Energy and Construction, Department of Trade and Industry, 1 Victoria Street, London SW1H 0ET, UK*

The Future of Today's Energy Sources

BERNARD J. BULKIN

1 The 1973 Energy Crisis

Crises inspire change. In 1973, the oil embargo associated with the *Yom Kippur* war led many people to look at the technologies around energy and predict that radical change would occur. There were forecasts, taken very seriously, that by 2000 oil would be nearly gone. The internal combustion engine was seen as having reached the limit of its ability to develop. Environmental concerns over air pollution were also seen as driving the rapid growth of nuclear power, and there was considerable effort being expended on commercialization of solar energy. Yet even at that time, some futurists warned that 'energy crises' had occurred before, and rarely had the predicted outcomes. Examples were the energy crisis associated with the lack of sites for new water wheels, and those created in many places by lack of wood through deforestation.

The main outcome of the 1973 'energy crisis' was efficiency. In the United States, as shown in Figure 1, the fuel economy of the car fleet doubled between 1973 and 1980. Appliances also became much more efficient: the average new refrigerator in the United States by the late 1990s was nearly 300% more efficient than in 1973 (Figure 2). These changes, which took place relatively rapidly, have shown great staying power.

The other big outcome was a shift away from oil as a fuel, particularly for power generation. In 10 years, oil went from being 17% of US power generation to about 2.5%. The shift was technically possible (coal and nuclear were the big winners, but once this driving force was removed, nuclear growth slowed) and economically desirable.

However, the core technologies of how we generate and consume energy, and the fuels we use, did not change radically. Perhaps the biggest change on the generation side, the growth of combined cycle gas turbines, was less a technological breakthrough than the conscientious application of basic principles of thermodynamics, made possible by a combination of materials science and information technology. As far as use goes, cars are still powered by internal combustion

Issues in Environmental Science and Technology, No. 19
Sustainability and Environmental Impact of Renewable Energy Sources

Figure 1 The fuel economy (in miles per US gallon) of new cars and light trucks sold in the United States from 1972 to 1998. Fuel economy of cars doubled during this period. The decline in the line showing both cars and trucks reflects the greater proportion of light trucks in new vehicle sales (source: US Department of Energy)

Figure 2 Efficiency of refrigerators sold in the United States, measured as the power required for a given volume, during the period 1972–1998. The pattern of steady improvement, taken to a limit by regulation, could be replicated in many other areas (source: US Department of Energy)

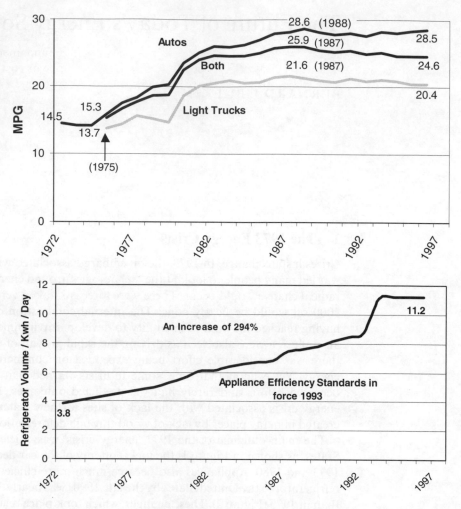

engines (though with increased use of another old technology, diesel engines, in Europe), not by batteries or Stirling engines. Refrigerators look and function in much the same way today as they did 25 years ago.

There are lessons to learn from a look back of a quarter century.[1] We see that radical change is possible. Doubling fuel economy of cars, tripling the efficiency of some major appliances, major changes in fuels (the UK went from 87% of its primary energy coming from coal in 1950 to 18% in 2001) do happen, and they are not fluctuations, but have long lasting effects. However, we also learn that the system has great resilience. There is a huge entrenched infrastructure which has great efficiency associated with it, and it takes major systemic shock (such as occurred in 1973) to upset it.

[1] A review of the 25th anniversary of the 1973 oil embargo and its consequences can be found at www.eia.doe.gov/emeu/25opec/anniversary.

There are also important lessons on supply. Predicting the future supply of any fossil fuel is probably impossible. Supply over the timescales we can consider relate to margin—price minus cost—rather than to fundamentals of the geology of the planet. It is also difficult to predict either price or cost with any degree of accuracy, so the uncertainty in the difference is very great. In subsequent sections this will be illustrated with reference to developing an understanding of the current reserves of the major fuels used today, and views about the future supplies of these fuels.

2   How Fossil Fuels Have Affected Our Lives   *history.*

Until the 19th century, human progress was limited by the amount of work that people could do in a day to feed themselves and their families. The economy was largely rural as a result. Beginning in the 19th century, people began to develop coal, oil and other stored energy sources to supplement solar energy. The results of plant and animal growth through solar energy, over huge areas and geologic time periods, coupled with violent geological upheaval, became available for human exploitation. Knowledge was required to develop machines capable of coupling these power inputs to human needs, and to be sure the great scientific work of the pioneers of thermodynamics and its application were critical. But what really changed the nature of how people lived on this planet was the several orders of magnitude increase of the energy sources that were available. There was, and continues to be, excess energy available to obtain more fossil fuels, to do research on how to exploit these fuels more efficiently, and to use them to drive change in cultures. Progress from that time until now has been 'like a flash explosion compared to the steady fires of the evolutionary record for previous millions of years'.[2]

One of the most important results of industrialization based on these energy sources is abundant food. Odum has pointed out[2] that it is an illusion, and a conceit of industrial society, that we are better at using the sun to grow food than our predecessors. Nothing could be further from the truth. In fact, we no longer eat food made completely from solar energy: we now eat potatoes made partly of oil! The same is true in the growth of animals for both meat and dairy products. This is not just the case on the farm, but also in the factories where farm machinery is made, fertilizer is manufactured, and in the universities where farming research is done on productivity. In effect, as Odum makes quantitatively clear, our society is developed around a 'fossil fuel subsidy', somewhat supplemented by a nuclear energy subsidy. In this paper, one of the questions we are looking at is: how long can this continue?

3   Energy Use in the World Today: Fuels and How They Are Used

Figure 3 shows a view of the energy use in the World. It is conveniently divided into four major categories: transport, power generation, industry/domestic/agri-

[2] H. T. Odum, *Environment, Power, and Society*, Wiley-Interscience, New York, 1971, pp. 104–138.

B. J. Bulkin

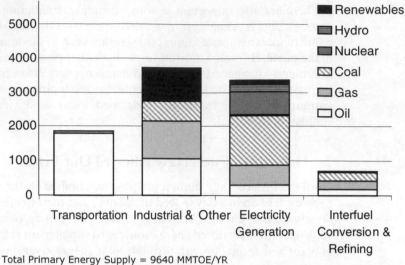

Figure 3 World primary energy supply, 1999 snapshot. The sector labelled 'Industrial and Other' includes domestic energy use and agriculture. The quantities shown are millions of tonnes of oil equivalent per year, which compares all fuels on a constant energy basis. For conversion factors between fuels, see www.bp.com/centres/energy

Total Primary Energy Supply = 9640 MMTOE/YR
Source: IEA World Energy Outlook. Data for 1999

culture, and the small amount used to convert one fuel to another (*e.g.* energy cost of refining of petroleum into gasoline and diesel fuel). The figure shows not only the quantities for each category of energy use, but also the fuel mix of the category.

As is well known, transport is completely dominated by oil. All other fuels combined are not sufficient to show up on a graph this size. Clearly the oil business is very dependent on demand from the transport sector, and the transport sector is very vulnerable to any disruption of supply of oil, as the entire infrastructure is built around a single fuel source.

This is not the case for the other major sectors. Power generation is very diversified, with coal, gas, nuclear, hydroelectric and oil all having significant shares. Renewables in this case are mainly wind power and a small amount of solar photovoltaic. In the case of power generation, the diversification of fuel can even occur locally; that is, a particular electric utility supplying a populous region might derive electricity from as many as four different fuels.

The energy sources for industrial/domestic and agriculture are also very diversified. Here renewable energy, which is a substantial contributor, is almost entirely burning of biomass in developing world countries. It is questionable whether this should be classified as renewable, as in many cases it results in deforestation and poor land use.

This chart and the trends that lead to the current situation are practically all one needs to know about today's energy sources. The following are key observations:

• Transport receives a huge amount of attention in regulation and technology development, but represents less than a quarter of all energy use. Growth in the developed world is very slow in this sector. Most vehicles are in cities, and congestion is limiting growth. In the United States and parts of Western

4

Europe there are already as many cars as licensed drivers. The potential for growth in China and India is still very great.

- Power generation is a big sector, and continues to grow strongly everywhere in the world. By contrast with transport, there seems to be nothing that will limit demand for electricity, as even the most affluent societies find new uses for electric power. As far as fuels for power generation are concerned, the last 30 years have seen significant reduction in the use of oil, and it is possible that the remaining oil could be squeezed out, certainly from all but small generators; despite widespread perception to the contrary, nuclear power continues to grow (2.8% in 2001), both through a few new plants and from higher utilization of existing plants; renewables are growing at a very rapid rate, especially wind and photovoltaics, but from a very small base and in small increments—new capacity is generally in kilowatt additions, in contrast to big power plants in the hundreds of megawatts range; there is a continuing shift towards gas and away from coal, for environmental reasons, but the coal reserves of China are much greater than the gas reserves, and one can expect a significant amount of China's future electric power to be coal generated.
- The largest sector, industry/domestic/agriculture, is the most diverse in every way. It is closely tied to GDP growth, so most affected by worldwide economic conditions. It is also the most distributed, and the hardest to regulate. While there is great diversity in fuel use, it is often difficult to effect change in fuels, as a particular user in a given geographic region is often tied to just one fuel. The greatest potential here is for efficiency improvements.
- Projections of how this chart will change over the coming two decades are extremely difficult and involve big assumptions. Figure 4 shows the evolution over the past 25 years, but may not be a guide to the future. For example, there are projections that are widely used showing that power generation will double from 2000 to 2020, and that the share of gas will also double (so an absolute increase in gas consumption for power of four times), while the share for nuclear power will halve. However, this assumes that no new nuclear power plants will be built during this period, and that some of the existing ones will not be relicensed to operate when their current licenses expire.

4   Nature of Oil, Coal and Gas *history*.

While we talk about the fossil fuels as if they were three substances, the reality is quite different. Each of these is found in the world with a wide variation in composition, and this variability is a big factor in determining the future of fossil fuels, both locally and globally.

In many ways, gas is the simplest. After all, methane is methane. However, as one might expect, natural gas resources have small but varying levels of C_2–C_4 hydrocarbons, hydrogen sulfide and, more importantly, most natural gas is found with CO_2 and nitrogen. The proportion of CO_2 varies widely, from less than 1% to more than 70%. While separation of the hydrogen sulfide is straightforward and relatively inexpensive, and the higher alkanes can be separated if there is an economic use for them, or left with the methane, the carbon dioxide poses more complex problems. It greatly affects the heating value of the fuel, as

5

Figure 4 World primary energy consumption, by fuel, from 1976 to 2001. The figure highlights the significant growth of gas and nuclear during this period. As in Figure 3, in millions of tonnes of oil equivalent (source: BP Statistical Review of World Energy, 2002)

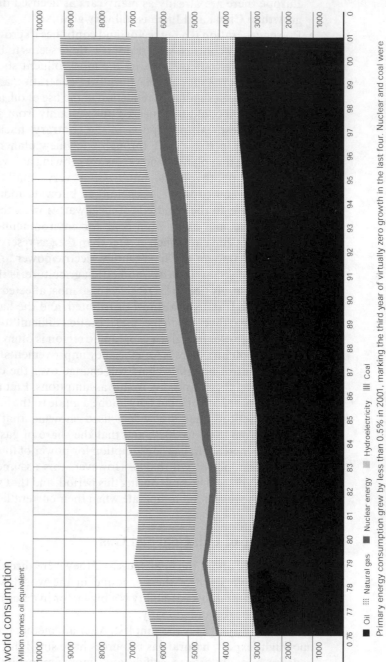

world consumption
Million tonnes oil equivalent

■ Oil ⠿ Natural gas ■ Nuclear energy ▦ Hydroelectricity ▥ Coal

Primary energy consumption grew by less than 0.5% in 2001, marking the third year of virtually zero growth in the last four. Nuclear and coal were the fastest-growing fuels, with hydroelectricity showing a steep fall. Coal increased its share of the overall energy market for only the second time since 1985.

Table 1 Composition of the main types of coals[a]

Type of coal	C (wt%)	H (wt%)	O (wt%)	N (wt%)	Moisture (wt%)	Calorific value (kJ/g)
Peat	45–60	3.5–6.8	20–45	0.75–3.0	70–90	17–22
Brown coals and lignites	60–75	4.5–5.5	17–35	0.75–2.1	30–50	28–30
Bituminous coals	75–92	4.0–5.6	3.0–20	0.75–2.0	1.0–20	29–37
Anthracites	92–95	2.9–4.0	2.0–3.0	0.5–2.0	1.5–3.5	36–37

[a]*Note*: all elemental percentages are given on a dry, mineral-free matter basis. Adapted from K. S. Vorres, *Kirk-Othmer Encyclopedia of Chemical Technology*, 4th edn., Wiley-Interscience, New York, 1995.

does the nitrogen. Separation of the CO_2 is costly, more so since there is little market for it. For giant fields with a high CO_2 content, such as the Natuna field in Indonesia, after separation the CO_2 must be reinjected or sequestered if it is not to have a major impact on atmospheric CO_2 levels.

Oil is more complex. On the one hand, crude oils are similar in that they contain some of every possible hydrocarbon isomer from C_4 up to very long chains. However, there the similarity ends. There is a very great variation in the chemical composition of crude oils found around the world. This manifests itself in very different viscosities, sulfur levels, percentages of alkenes, aromatics, naphthenic and asphaltenic hydrocarbons, heavy metals, *etc*. All of this arises from variations in the conditions of formation and trapping of the oil. The lightest crude oils have viscosities similar to water or light hydrocarbons; the heaviest will not flow even at fairly elevated temperatures. The relevance of this for the future supply of oil will become clear shortly.

Coals also vary in chemical composition and structure. Again, this depends on the conditions of formation of the coals, and also on their age. Some types of coal and their chemical composition are given in Table 1.

Why is this variation in composition of the fossil fuels relevant to a discussion about their future? Because today we make use of only those that can be processed into useable fuels most cheaply. Any beginnings of depletion of the central resource being used today would bring into play new technologies for exploiting further resources. Indeed, anything that either raises price or lowers costs, independent of worries about depletion, brings new resources to market. So today, world oil reserves are 15 times greater than they were when record keeping began in 1948; world gas reserves are four times greater than they were 30 years ago, and world coal reserves have risen 75% in the last 20 years.[3] However, for example in the case of oil, this only takes into account what is arbitrarily classified as conventional oil.

Petroleum is a broad term that includes mixtures of hydrocarbons ranging, in some cases, from gases such methane, ethane and propane, to fluid light oils, through to more viscous heavy oils, and on to shales, tar sands and bitumens. While some gas, often large quantities of gas, is almost always found associated

[3] R. L. Bradley, *Policy Analysis*, 1999, available at www.cato.org/pubs/pas/pa-341es.

Figure 5 The magnitude of conventional oil and gas reserves (source: US Geological Survey) compared to the quantities of extra heavy oil reserves in three main basins, Olenek in Russia, Athabasca in Canada and Orinoco in Venezuela. The quantities shown are billions of barrels of oil equivalent. Note that 1 barrel of oil is 159 litres

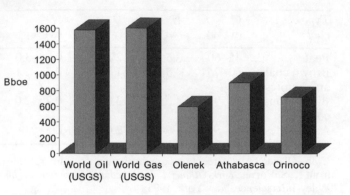

with oil, there are also large gas fields that have smaller amounts of liquids associated with them. These liquids, usually the lighter end of the molecular weight and viscosity range, are known as Natural Gas Liquids or NGLs. Today, more than 95% of the oils that are produced are light fluid oils, that flow at room temperature or moderately higher temperatures above ambient. Some of this oil is mobilized to flow by using gas under pressure or water flooding. These liquid oils, and NGLs, are together known as conventional oil. Everything else—heavy and extra heavy oil, tar sands, shales and bitumens—that need to be mined and treated before they can be processed further, are known as unconventional oil. We see that there is a spectrum of materials found in Nature, from gas to very heavy hydrocarbons. The materials, besides varying in molecular weight distribution and viscosity, also have considerable variation in hydrocarbon type (as mentioned above, alkanes, alkenes, aromatics, *etc.*), sulfur content and metals.

On the higher average molecular weight side are a range of resources known as heavy oil or extra heavy oil. The amounts are huge. It is possible that Venezuelan extra heavy oil could be as much as the world's total conventional oil reserves. More conservative estimates say that Canada and Venezuela have extra heavy oil equal to the world total for conventional oil. Figure 5 shows the amount of world reserves of conventional oil and gas (as oil equivalent) compared with three large heavy oil basins in Russia, Canada and Venezuela. While estimates vary considerably, the total unconventional oil resources are accepted to be about three times the volume of conventional oil that there was originally, before production, and about 10 times the volume of recoverable conventional oil that remains.

Technology to recover this unconventional oil continues to improve. It needs to be mined, rather than pumped, or liquefied by steam and chemically modified. There are often very high sulfur levels and high heavy metal contents that need to be removed before the oil can be processed in a refinery. All of the production is energy intensive, so the ratio of energy coming out to that going in is much lower than for conventional oil, though still higher than some alternatives that are being advocated. This is true of costs as well: while higher than conventional oil, even today they are lower than the costs of many alternatives, especially from biomass. We can expect to see steady, albeit incremental, improvement in the technology to process all unconventional oils.

5 Reserves of Fossil Fuels

Figure 6 summarizes the reserves of oil in the world, showing the geographic distribution (heavily concentrated in the Middle East) and the evolution with time over the past 20 years. While there has been some criticism of the validity of the growth of Middle Eastern reserves depicted in the figure, it seems clear that proved reserves of oil are now as high as they have ever been, despite consumption also being higher than it has ever been.

Figure 7 shows the equivalent data for natural gas. Here we see that the big reserve concentrations are in the Middle East and Former Soviet Union, and that the growth of reserves is even more dramatic than in the case of oil. Most exploration companies believe that there is potential for further growth in gas reserves, as gas becomes an important product globally. To date, much of the exploration effort has been concentrated on finding oil, and gas has been sometimes considered as a failure of oil exploration.

In energy terms, coal reserves are enormous compared to either oil or gas. While, at today's consumption levels, oil reserves are sufficient for *ca.* 40 years, and gas for *ca.* 60 years, coal reserves would suffice for more than 200 years (Figure 8). By contrast, little coal is found in the Middle East, but there are large reserves in Asia, North America, the Former Soviet Union and Europe.

The point of all of the foregoing discussion is this: change may be coming in our energy mix, and how it is used. However, this change, at least for the coming several decades, will not be driven by depletion of existing fuels. To make a case for change coming, one needs to look beyond reserves. That means we must examine factors related to getting these reserves to markets, and the demand of the markets.

6 The Future of Demand

In the introduction, we have shown how efficiency of the vehicle fleet changed during the 1970s, and how appliances increased in energy efficiency. Could the demand for today's energy sources be further influenced by efficiency? What about the elimination of waste?

Figure 9 is a complex figure showing the flow of energy through the US, indicating the efficiency of each sector. Of course, thermodynamic efficiency will always be less than 100%, and there are some losses that are inevitable (friction losses of tyres on roads cannot logically be reduced to zero!), but it is apparent that there are still big opportunities for gains, despite what has already been accomplished. It is generally agreed that the situation in Europe is not dissimilar, and that the rest of the world is worse. Indeed, while the US is a big (and some would say, disproportionate) energy consumer, it does so with relatively good efficiency. Nonetheless, if we take the ratio of useful/(useful + rejected) energy as a crude measure of efficiency, we see that transport is 20% efficient, power generation 32%, while the direct inputs of natural gas, coal, oil and biomass into residential, commercial and industrial use are more efficient.

It is this low efficiency, no matter how measured, that has focused attention on the transport and power generation sectors. The future of demand for today's

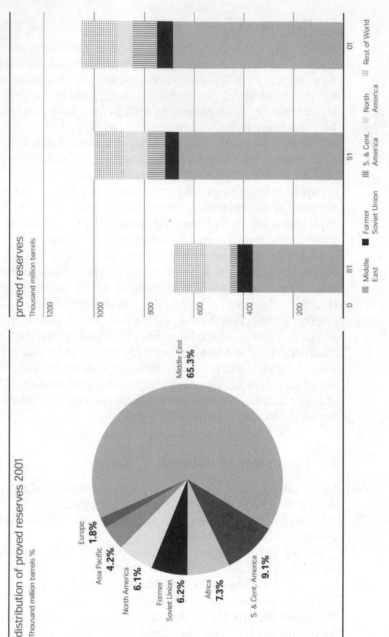

Figure 6 Proved reserves of conventional oil, showing that *ca.* 2/3 of the oil is in the Middle East, as well as how reserves have grown over the past two decades (source: BP Statistical Review of World Energy, 2002)

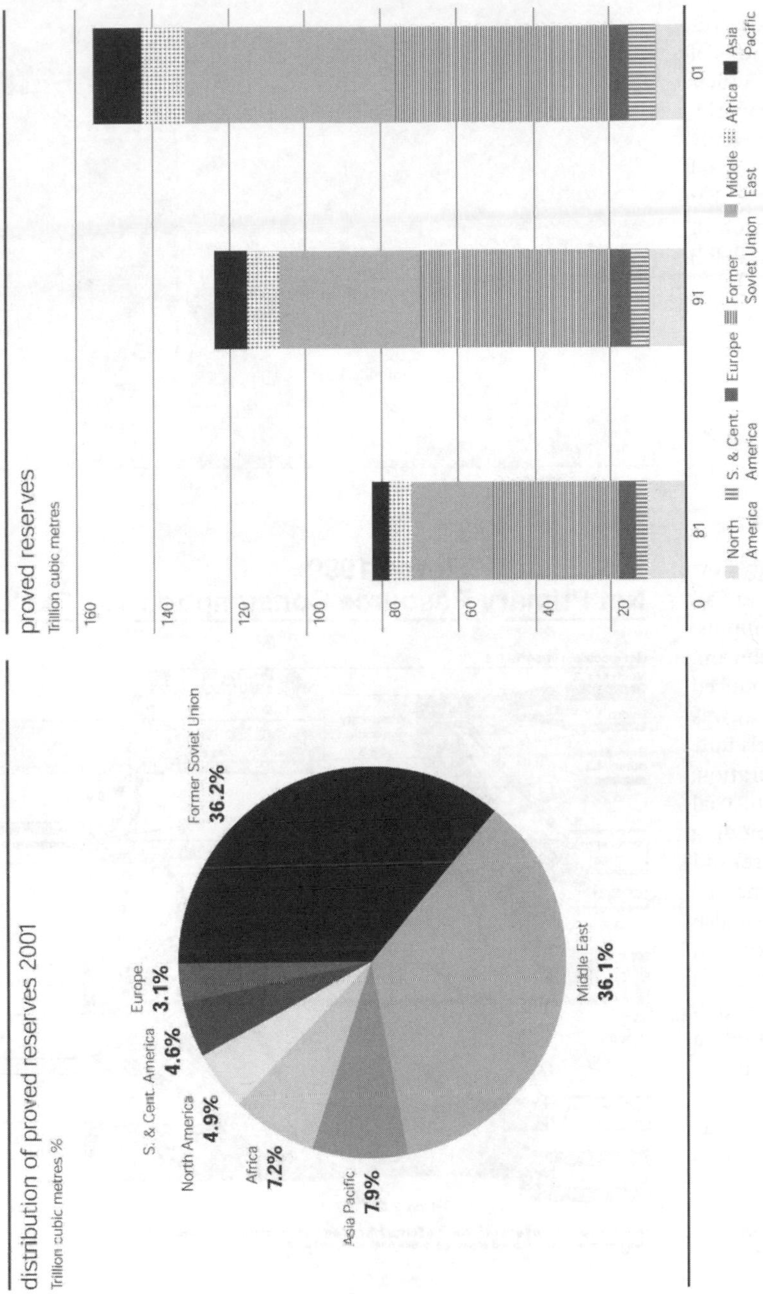

Figure 7 Proved reserves of natural gas, in trillions of cubic metres, showing the importance of the former Soviet Union, as well as how reserves have grown over the past two decades (source: BP Statistical Review of World Energy, 2002). The web site given in the caption to Figure 3 contains conversion factors to oil equivalents and the energy available from these reserves

Figure 8 Proved reserves of coal (in billions of tonnes of coal) and their distribution in the world. Coal is more evenly distributed than either oil or gas (source: BP Statistical Review of World Energy, 2002)

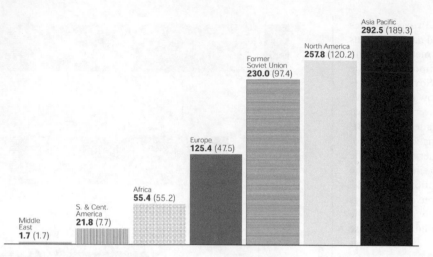

Figure 9 US energy flow in 1999. On the left are all the fuels making inputs, showing, where relevant, what was produced domestically and what was imported. The fuels flow into power generation, transport, industrial production (including agriculture) and residential/commercial uses. Power generated also flows into industry and residential/commercial uses. The map then shows the amounts (all in exajoules) that do useful work and the amounts that are wasted (source: US Department of Energy)

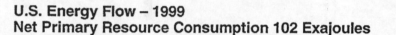

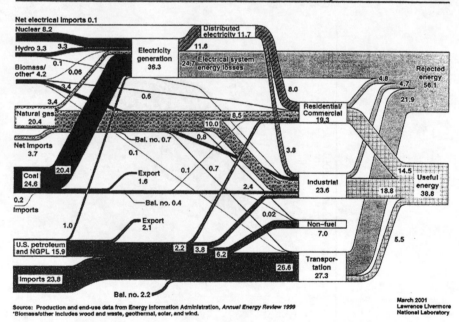

energy sources depends on what happens to this efficiency. Could the dramatic changes in efficiency that took place in the last 25 years be repeated?

Transport

The easiest sector to look at is road transport. Clearly there are several variables: the fuel economy of the new vehicles being sold (including a minimum fuel

Figure 10 Schematic view of a series hybrid vehicle. The wheels, at right, are driven by electric motors, drawing power from a battery. The battery is charged *via* an internal combustion engine. Energy from braking is captured *via* a regenerative braking system

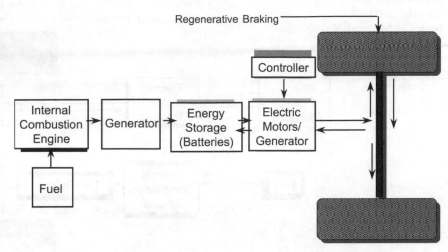

economy requirement and corporate average, if any), the mix of high and low fuel economy vehicles actually sold (which has dominated the last decade with the growth of light trucks as private passenger vehicles), the rate of turnover of the fleet (which varies greatly from country to country), driver behaviour and traffic management. Added to this is consideration of what is technically feasible.

At one extreme, we have current reality, that is, a fleet average for cars of 25–30 miles per gallon or *ca.* 8 L/100 km. At the other extreme, Volkswagen has built a two-seater car that has been driven on German highways, achieving just over 1 L/100 km. Lovins,[4] in several publications, has hypothesized a car achieving 0.5 L/100 km. This radical improvement in fuel economy is achieved even with internal combustion engines, using a large number of known technologies. Crucial is weight reduction, achieved through light materials. As Lovins points out, in today's car, 95% of the useful energy moves the car, and only 5% moves the driver, in proportion to their weight. Aerodynamic design to minimize loss also plays an important role. As fuel economy is improved through optimizing these variables, as well as the engine size and performance, the quantity of fuel stored on board the vehicle can be reduced to allow for a constant range between refuelling. This further reduces weight.

Moving beyond optimized internal combustion (IC) engines, hybrid vehicles appear to be very promising for increasing the efficiency of transport. Hybrids combine an IC engine with an electric drive, either in a series (Figure 10) or parallel (Figure 11) configuration. Hybrids readily achieve improvements in fuel economy of 2–3 times that of a comparable IC vehicle. They do this through several means, relying on the fact that the least efficient use of an internal

⁴ P. Hawken, A. Lovins, L. H. Lovins, *Natural Capitalism*, Little Brown, Boston, 1999, pp. 22–47 and references therein.

Figure 11 Schematic view of a parallel hybrid vehicle. Comparing with Figure 10, in this case the wheels can be driven by either the electric system or by the internal combustion engine, the latter also being available to charge the batteries

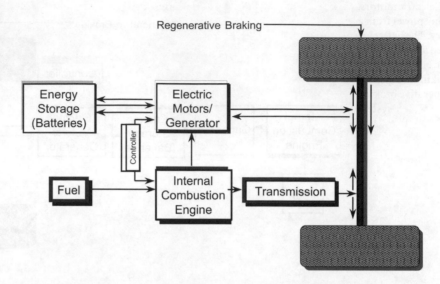

combustion engine is in acceleration and deceleration. They use the efficiency of electric motors as drivers of the wheels (and the fact that electric motors do not consume fuel when they are stationary); they capture energy of braking through a technique known as regenerative braking (effectively running a generator while braking), and they use a smaller internal combustion engine running in its most efficient range. When the vehicle is accelerating, it can draw down on the batteries to achieve extra power. Unlike an electric vehicle, it does not need to rely on a large battery pack for range.

The obvious area in which hybrids will penetrate transport is the urban bus. Urban buses are always accelerating or decelerating, rarely running at constant speed. Therefore, the benefit achieved by a hybrid is maximized. Several urban transit authorities in New York and California have evaluated hybrid buses, using low-sulfur diesel as the fuel for the internal combustion engine. The hybrid was compared with other alternatives, such as compressed natural gas (CNG) or liquefied petroleum gas (LPG) as the fuel. In all cases the hybrid was the preferred solution. Its emissions were comparable on most parameters to those of CNG or LPG, but there was no infrastructure cost for converting the fuelling system. Because of the lower cost, more hybrid buses could be brought into service for the same expenditure, and the air quality impact was much greater than for alternatives.

There are now several hybrid passenger car models on the commercial market, and more are expected in the near future. They are more expensive than a simple internal combustion engine, of course, as there are effectively two separate means of powering the vehicle. Nonetheless, they are expected to appeal to several segments of the market.

Hybrids are particularly relevant to this chapter, as they represent a way of using today's energy sources (in this case, oil converted to gasoline and diesel) in a far more efficient way. They thus raise the barrier for all alternatives to gasoline and diesel to make headway. Except for particular niches, it is probably impossible for alternatives such as CNG to overcome this barrier. There is little technological stretch in these alternatives that will make them more competitive over time. Fuel cell powered vehicles still have a long way to go technically. They do offer greater efficiency and lower emissions than hybrids, and represent the only realistic long term challenge to gasoline and diesel fuel. Nonetheless, all hydrogen for fuel cells for the coming decades is likely to be generated from fossil fuels—mostly from natural gas, but with the possibility of coal-generated hydrogen being a realistic one as well. Again, the infrastructure costs for hydrogen fuelling are high.

Power Generation

We have already seen that the trends in power generation over the past 25 years are:

- Growth in demand, expected to continue without much constraint; world demand for electricity projected to grow at 2.6% per annum for the next two decades
- Growth in nuclear power, but not as fast as predicted, because of fears of danger, fears of proliferation, and inability to deal with waste
- Growth in the use of natural gas, and in the efficiency with which it is used, with combined cycle gas turbines becoming a dominant technology
- Levelling off of the use of coal, with either low-sulfur coal or scrubbing techniques becoming dominant in the US and Europe, and with coal being a higher cost than gas for new plants in most locations
- Reduction in the use of oil, likely to continue
- Rapid growth in renewables, especially wind and solar, but from a very small base, so still remaining as a very small share of the total fuel mix

The most likely scenario for the future is that fossil fuels will continue to provide most of the energy for the increased demand, unless a combination of government policy and public attitudes change the view of nuclear power.

The environmental concerns associated with the growth of electricity generated by fossil fuel are numerous, but at this time are dominated by particulate matter, especially fine particles, and their precursors (oxides of nitrogen and sulfur; trace element emissions, especially metals; and carbon dioxide).

We have seen that efficiency of power generation worldwide is *ca.* 35%, dominated by coal-fired power plants achieving efficiencies in this range. Modern gas combined cycle plants exceed 50%, but we have already stressed that this is far from the thermodynamic limit.

The future potential is perhaps best estimated by looking at the goals of the US Department of Energy Vision 21 program,[5] an extensive academic/industrial/government program of research. Their efficiency goal is 60% for coal-based systems (based on higher heating value) and 75% for natural gas-based

---

[5] A description of the Vision 21 programme can be found at www.fetc.doe.gov/coalpower/vision21.

systems. This translates, for the fuel mix, to a 40–50% reduction in greenhouse gas emissions from power generation. This is combined with aggressive targets for reduction of particulates, nitrogen and sulfur oxides, and metals.

The technology roadmap to achieve this involves a large number of distinct areas of improvement and innovation, building on developments in information technology and materials science, as well as a greatly improved understanding of combustion processes.

The fundamental science elements include:

- Computational modelling and simulation at the plant level
- Development of new sensors and controls for much higher quality plant operation
- Systems integration of all the new technologies and concepts
- Advanced materials such as ceramics for turbines, membranes for gas separation, and hot gas filter materials

They are also examining nitrogen-free combustion with membrane-based air separation, re-examining coal gasification, purifying exhaust gases for eventual sequestration of the carbon dioxide, and looking at the integration of fuel/chemical co-production from coal with power generation. An applied program also looks at the scale-up of high-efficiency fuel flexible turbines. Fuel cells are a longer-term part of the program as well, and fuel cell–turbine hybrid systems are part of this vision.

The important thing about this program of improving efficiency of power generation, using coal and gas as the fuel sources, is that the benefits come from a large number of incremental innovations and developments that can be implemented independently. We can reliably expect to see continuous improvement in efficiency and reduction of emissions over a 20-year period.

7   Sequestration

No amount of efficiency improvement will eliminate all of the carbon dioxide produced in fossil fuel combustion. Far from it. Switching to zero carbon (*e.g.* solar, wind) technologies, or even increasing nuclear power generation, will happen over a long period of time. So the final important element of a strategy for maintaining the viability of fossil fuels for the coming decades is carbon sequestration.

The term 'carbon sequestration' has been used to include several different technologies. It means taking carbon dioxide that has been emitted and storing it in some way that does not increase the concentration of CO_2 in the atmosphere. The big technological options for carbon sequestration thus involve separating CO_2 from other gases at the source, capturing the CO_2 in a form suitable for storage, and storing it or reusing it. Storage would mean injection into non-atmospheric sinks such as depleted oil and gas reservoirs (where it might replace methane that had already been used), coal seams, deep saline formations or the deep ocean.

Sequestration has also been used to refer to other 'carbon sinks' such as forests or soils, where the separation and capture of the CO_2 is not a part of the process, but the sinks take up more CO_2 in enhanced natural processes, effectively re-equilibrating from an increase in atmospheric concentration. Because the ocean is the largest natural sink for CO_2, this can also apply to enhanced take-up of CO_2 into the oceans, as distinct from CO_2 capture and injection.

A good overview of all carbon sequestration technologies is available from Herzog *et al.*[6] A more recent paper by Herzog[7] describes the technology for sequestration in oil reservoirs in more detail, with reference to the one experiment at scale that has been carried out in Norway.

Herzog has pointed out that the potential for sequestration into the ocean is huge: the oceans already contain huge quantities of carbon that has been sequestered by natural processes, mostly as carbonates, and all of the carbon that is involved in a doubling of the atmospheric concentration of CO_2 would only change the ocean carbon content by 2%. However, this proposal is extremely controversial. Environmental groups have pointed out that the effects on life in the deep oceans would probably be considerable. Using the oceans as a disposal site is probably not acceptable to most of the population today.

By contrast, geological sinks (such as oil reservoirs, coal seams and saline formations), while having an order of magnitude smaller capacity, offer many advantages. In general there is little if any biological life in such places. Also, there can be economic benefits that help offset the cost of the sequestration. CO_2 is known to displace methane from coal surfaces, so with proper management injected CO_2 could produce a stream of methane that would be revenue generating. CO_2 is already used to enhance recovery of oil from reservoirs, and it seems likely that there is considerably more potential for doing this.

However, there are technical challenges. The CO_2, for example from power plants, is generally a mixture with other gases, many of them far from benign (such as oxides of sulfur and nitrogen), so gas separation technologies need to be improved. Work needs to be done on demonstrating that injected CO_2 stays where it is put, over long periods of time. So leakage rates need to be calculated, measured and demonstrated to be negligible and not subject to catastrophic failure. Finally, costs, while not impossibly high, probably need to be reduced by an order of magnitude if sequestration is to be brought into widespread use.

The capacity of forests to sequester CO_2 is much smaller. There is not sufficient capacity in the biosphere to sequester the carbon from the lithosphere. However, afforestation and reforestation have appeal, and offer other local benefits. In Western Australia, for example, the state has used reforestation to help control soil salinity, improve the appearance and soil retention of farms, while storing carbon. It is probably only through these sorts of carbon sinks that one can deal with CO_2 from dispersed sources, such as transport vehicles or homes.

Carbon sequestration in whatever form does not offer a magical solution to the problems of climate change and fossil fuels. Nevertheless, it probably represents a piece of the solution, when combined with lower carbon fuels (such as substituting methane for coal) and higher efficiency, that makes a fossil fuel future more sustainable, or prolongs the transition to alternatives.

[6] H. Herzog, B. Eliasson and O. Kaarstad, *Sci. Am.*, 2000, **282**(2), 72.

[7] H. Herzog, *Environ. Sci. Technol.*, 2001, **35**, 148–153.

Sustainable Energy: Choices, Problems and Opportunities

DAVID ELLIOTT

1 Introduction

The development of patterns of energy generation and use which can be sustained into the future is increasingly being seen as urgent, given growing concerns about the potential social and economic impacts of climate change. This article reviews the energy options for a sustainable future, focusing on options which might reduce or limit the level of carbon dioxide emissions.

The climate change issue is a global one and, increasingly, developing countries will begin to contribute to it as they industrialize. However, the industrialized countries bear the responsibility for having developed the technologies that have created the problem. It would only be reasonable to expect them to play a leading role in developing some of the technologies which might help reduce it.

While the overall view is global, this article draws on examples of developments chiefly from the UK, which has embarked on a major programme of sustainable energy development. That is not to suggest that there are not other developments going on elsewhere; indeed, the UK is in many respects following behind initiatives pioneered by countries like Denmark and Germany, particularly in the wind power field.

The main focus of this review is on the new renewable energy technologies, since this is the main topic area of this book. However, to set the scene, the prospects for the other main options, including carbon sequestration, energy conservation and nuclear power, are reviewed. Subsequent sections then attempt to make a comparison of the merits of each option, so as to provide an overview of the choices ahead.

The basic criteria for the energy choices facing the world are, of course, not just environmental. Economic factors remain crucial. However, short term economic prosperity may mean little if the health, well being and livelihood of the populations concerned cannot be maintained. At one time a major pre-occupation of many environmentalists was the fear that supplies of key fuels would soon be

Issues in Environmental Science and Technology, No. 19
Sustainability and Environmental Impact of Renewable Energy Sources

exhausted. Nowadays, although the price of fuels remains a key political issue, fuel scarcity is less of an immediate concern. Instead, the main concern is whether we can safely use the reserves of fuel we have left.[1]

2   Climate Change

The scale of fossil fuel combustion has increased dramatically since the industrial revolution, by a factor of more than 20, and unless changes are made to the way energy is generated and used, this growth seems likely to continue. World energy consumption is projected to increase by 59% from1999 to 2020, with fossil fuels providing the bulk of this energy.[2] In the industrialized countries, their use is split roughly equally between electricity generation (coal), heating (natural gas) and transport (oil), although gas is increasingly replacing coal for electricity generation and the transport sector is expanding continually.

However, the use of fossil fuels creates environmental problems. In addition to problems with acid emissions, the combustion of fossil fuels in power stations and vehicle engines inevitably generates carbon dioxide gas, and these emissions, along with other greenhouse gases like methane, are claimed to play a significant part in the process of global climate change that seems to be underway. While there are few who now doubt that climate change is a reality, not everyone agrees that it is necessarily the result of human activities, or that we should and could do something about it. However, the Intergovernmental Panel on Climate Change (IPCC), representing the bulk of the world's scientists, has concluded that human activities are the main cause and that, unless action is taken, we can expect increasing ecological, social and economic problems around the world. As weather patterns become more erratic, there are likely to be increasingly violent storms and floods and at other times droughts, and as average temperatures rise, sea levels will rise, with low-lying areas in the world being inundated.[3]

It is hard to put accurate figures on the economic cost of the potential damage and dislocation, not least since much of the impact of climate change could fall on developing countries in terms not just of loss of food growing areas, but also increased disease and social disruption. However, just focusing on the industrial countries, the insurance company Munich Re has estimated that if carbon emissions doubled then that would impose direct costs on the UK of around $25 bn pa and $150 bn pa for the USA, which is 1.4% of the USA's current GDP. This assessment is likely to understate the full magnitude of the global problem. The former UK Government Chief Scientific Advisor, Sir Robert May, has noted that, in addition to the costs of the direct damage to property and the economy, there could be major impacts on natural ecological processes such as soil formation, water supplies, nutrient cycling, waste processing and pollination, which would have indirect economic implications around the world. He reported rough estimates of the economic value of these processes as being around £10–34

[1] D. Elliott and A. Clarke, Developing criteria for sustainable energy technology, *Int. J. Global Energy Issues*, 1997, **9**, 264–274.

[2] *World Energy Outlook*, International Energy Agency, Paris, 2001.

[3] Intergovermental Panel on Climate Change, third assessment report WG II, *Climate Change 2001: Impacts, Adaptation and Vulnerability*, Geneva, 2001.

trillion per year, about twice the conventional global GNP, and noted that *'large swathes of this £10–34 trillion are at risk from the possible environmental and ecological changes sketched by the IPCC'*.[4]

Some of this damage seems likely to occur whatever is now done, since the carbon dioxide imbalance that has already been created is likely to persist for some while (although there are continual interchanges between the air, the sea and the other carbon sinks, the net excess carbon dioxide can take many decades to be absorbed). Moreover, unless radical changes are made, the situation could be made even worse, as demand for energy increases. World CO_2 emissions are projected to rise from 6.1 billion metric tonnes of carbon equivalent in 1999 to 9.8 billion metric tons in 2020, a 60% increase, possibly rising to 15.1 billion metric tonnes by 2050, a 152% increase on 1990 levels.[2]

By contrast, the IPCC has recommended that, to try to avoid the worst excesses of climate change, the carbon dioxide level in the upper atmosphere should be limited to around 550 parts per million by volume, which is about twice pre-industrial levels. The current level is around 363 ppmv. This implies that a 60% reduction in emissions (from 1999 levels) should be made globally, a target which has also been recommended as relevant for the UK by the Royal Commission on Environmental Pollution, which also proposed a target date of 2050.[5] Given that nearly 80% of the energy generated around the world at present comes from fossil fuels, making a reduction on this scale will be difficult.

3 Carbon Sequestration

One way to try to deal with the problem of carbon dioxide emissions is to collect the gas and store it safely somewhere. Clearly this is not realistic for most vehicle exhausts, but it might be possible for power stations. However, it is far from cheap to capture carbon dioxide that would otherwise vent up the power station chimney. Estimates suggested that it might add 50–80% to the cost of electricity, and could reduce the efficiency of energy conversion by around 10%.[3] Moreover, finding somewhere to store the large volumes of gas is far from easy. The most promising option is to store it as compressed gas in depleted oil and gas wells.

There is a symmetry in this 'carbon sequestration' concept, since that is where at least some of the carbon originally came from. There are also situations where CO_2 gas injection can be used to improve the efficiency of gas or oil extraction before the wells are empty. However, overall there may only be space in empty wells for a few decades of emissions. Estimates of storage capacity range from 40 to 100 gigatonnes for oil wells and from 90 to 400 gigatonnes for gas wells.[6] Given that in 1999 some 6 gigatonnes of carbon dioxide was emitted worldwide, and emission levels are increasing, this storage capacity could only be sufficient for at most 80 years worth of emissions.

[4] R. May, *Climate Change*, UK Government Chief Scientific Adviser, 1997.

[5] Royal Commission on Environmental Pollution, *Energy—The Changing Climate*, 22nd report, Cmd 4749, London, HMSO, 2000.

[6] R. H. Williams, A technological strategy for making fossil fuels environment and climate friendly, *World Energy Council J.*, 1998, July, 59–67.

There could, however, be much more space in some saline aquifers, although there are relatively few 'closed' or sealed aquifers suitable for secure storage, possibly sufficient for around 50 gigatonnes. By contrast, there is much more room with less secure 'open' aquifers, possibly enough for up to 13 000 gigatonnes, or around 2000 years worth of current emissions. However, there is the problem of ensuring that the gas does not escape at some point owing to geological shifts, and there is a range of other environmental unknowns.[7]

Another somewhat more conventional sequestration option is biological storage in biomass, which absorbs carbon dioxide as it grows. For example, there have been moves to plant trees to compensate for carbon emissions from vehicles. However, there is not room to store all the emissions in this way. For example, to sequester all the UK's continuing carbon emissions in trees would require new forests to be planted over an area the size of Devon and Cornwall every year.[8] In addition, the storage is not permanent. Although some of the carbon may remain trapped in the roots, forests can burn and trees eventually decay, releasing the stored carbon dioxide.

It is usually argued that it would be more effective in carbon mitigation terms to plant and then rapidly harvest fast-growing, short-rotation coppices of willow, and use these energy crops to produce electricity in a power plant, assuming the energy produced substituted for energy that would otherwise have been supplied using fossil fuels. Nevertheless, reafforestation has its attractions as interim carbon store, since it is relatively cheap, assuming land is available, and it offers other benefits, such as enhanced biodiversity.

Forestry is only one option for creating new biological carbon sinks. The adopting of different farming practices can also offer routes to carbon sequestration. For example, a 'no till' policy can enhance the absorption of carbon dioxide by the soil. There is also the newly emerging, but very speculative, option of ocean sequestration. In addition to the idea of pumping carbon dioxide into the depths, there is some interest in the idea of enhancing sequestration in the oceans by seeding areas with iron oxide. However, modifying the ecology of large areas of the sea could create a range of environmental problems and this approach has been seen as likely to be expensive and inefficient.[5]

Leaving ocean sequestration aside, overall, it has been estimated that, globally, the various types of land based carbon sinks (trees, soil) could at best absorb about 100 petagrams of carbon emissions per annum, which is 25% of the 400 petagrams of carbon emissions that would have to be offset in order to meet the 60% overall emission reduction target by 2050 mentioned above. Within this context, sequestration in soil and in trees are seen as about equally viable in carbon mitigation terms.[9]

In the end, despite its attractions, carbon sequestration might be seen as a rather inelegant approach to dealing with the problem of emissions—essentially

---

[7] UN/World Energy Council, *World Energy Assessment: Energy and the Challenge of Sustainability*, UN Development Programme, UN Department of Economic and Social Affairs and the World Energy Council, 2000.

[8] Climate Care (the Carbon Storage Trust), quoted in *Green Futures*, July/Aug 2001.

[9] Royal Society, *The Role of Land Carbon Sequestration in Mitigating Global Climate Change*, Royal Society, London, July 2001.

trying to deal with the problem after the event. Surely it would be better not to produce so much carbon dioxide in the first place.

4 Energy Efficient Generation

The most direct way to reduce carbon dioxide production is by burning less fossil fuel. There are various ways in which this can be achieved. The simplest is for people to actually use less energy, *i.e.* to make do with lowered energy services, which implies something of a frugal approach to lifestyles. While there may well be some benefit in turning off lights when not needed, donning a pullover rather than switching on more heating, or voluntarily forgoing some particularly energy-intensive activity like flying, in general 'energy conservation' (or more accurately fuel conservation) can also be achieved by technical measures which reduce the waste of energy.

It is possible to use fossil fuels more efficiently, so as to get more useful energy for each tonne of carbon dioxide produced, or to put it more positively, get the same amount of useful energy with less carbon emissions. This can apply both to generation and consumption. The classic example in terms of generation is combined heat and power (CHP), the co-generation of heat as well as electricity in power plants. Traditional coal-fired power plants typically reject 70% or more of the energy in the fuel they use into the environment, mostly radiated out as heat from cooling towers. About half of this heat can be reclaimed and used to feed district heating networks. That can double the overall efficiency of energy conversion, or allow plants to deliver the same amount of total power from about half as much fuel, with proportionate decreases in carbon emissions.

CHP technology has been adopted quite widely in some parts of the world, notably in Northern Europe, but in recent years, instead of large coal fired CHP plants for city-wide heating, the emphasis has moved to much smaller gas fired units: micro CHP. Overall, the UK government's target is for there to be 10 GW of electricity generation capacity from CHP in place by 2010, in effect saving nearly 5 GW's worth of fossil fuel use, with much of this new capacity being at the small to medium scale.

Worthwhile as CHP is, the electrical power industry has mostly preferred another option: switching over to medium-sized combined cycle gas turbines (CCGT), without CHP. The reasons for their popularity is that gas has been cheap and gas-fired turbines are quick to install: essentially they are jet engines driving a generator, with the hot exhaust gases also used to produce steam for a conventional turbine, in a two-stage system. They are not as efficient as CHP, but are much better than conventional plants, achieving overall energy conversion efficiencies of 55% or more.

As a result of this increased efficiency, and since burning natural gas (methane) rather than coal results in lower carbon dioxide production per kWh generated, the use of CCGTs produces about 40% less carbon dioxide per kWh of energy generated than burning coal in conventional plants. Switching over from burning coal to burning gas to generate electricity has meant that the UK has been able to reduce its carbon dioxide emissions to keep in line with the international targets agreed following the Rio Earth Summit in 1992, without having to do anything much else.

23

The scale of this 'dash for gas' in the UK has been remarkable. Whereas 80% of the UK's electricity used to come from burning coal, now it is down to 28%, and gas provides 40%. However, there is a limit to how many more coal-fired plants can be replaced by gas plants, so the growth in emissions savings will tail off. As demand for power increases, as seems likely, then once again net emissions will increase. There are also limits to gas reserves. For example, with the North Sea reserves becoming depleted and expensive, then by 2020, on current trends, the UK could be getting 70–80% of its gas from remote regions in the north east of Russia or, if that proves difficult, from countries like Algeria. Given the geo-political uncertainties, not everyone is convinced that this would be a reliable scenario in terms of ensuring security of supply.[10]

There has been some interest in trying to clean up emissions from coal combustion, given that around the world coal reserves are more extensive than those for gas. There are two main technologies. Fluidized bed coal combustion plants are more efficient in burning coal than conventional coal plants and can reduce acid emissions. With integrated gasification combined cycle technology (IGCC), coal is converted into gas which is then burnt in a CCGT. However, these as yet only partially developed technologies are likely to be expensive, with capital costs/kW for fluidized bed plants perhaps twice that for CCGT, and even more in the case of IGCC. Moreover, the overall energy conversion efficiencies of these plants seem likely to be lower than for gas-fired CCGT plants, possibly only around 40% compared with 55% and higher for CCGTs. So they will produce more carbon dioxide per kWh generated.[11]

With the IGCC, there is the advantage that it is easier to collect the CO_2 from the coal gasification process than from the emissions of conventional coal combustion plants, so sequestration could be easier. However, it is unclear whether that would compensate for the extra cost and lower efficiency of the plant. A DTI report on cleaner coal technology (CCT), published in 2001, concluded that there was 'no economic case for building new coal plant . . . at present electricity prices. Even without carbon capture and storage, new CCT plant would produce electricity in the range 2.6–3.7 p/kWh; with carbon capture and storage, this would rise to something like 4.8–5.8 p/kWh'.[12]

5 End Use Energy Efficiency

While the use of gas and CCGT clearly has benefits, and is the dominant new energy option in many countries at present, given its reliance on an finite fossil fuel, this approach may only offer a temporary and partial solution to the emissions problem. By contrast, energy conservation at the point of use would

[10] House of Commons Select Committee on Trade and Industry, *Security of Energy Supply*, session 2001–2, second report, HMSO, London.

[11] J. Watson, evidence to the Government's Review of Energy Sources for Power Generation on Advanced Fossil-Fuel Technologies for the UK Power Industry, 1998; *Financing Cleaner Coal-Fired Power Plants in the UK*, SPRU report no. 20, Energy Programme of the Science Policy Research Unit at Sussex University, 1998.

[12] DTI, *Review into the Case for Government Support for Cleaner Coal Technology Demonstration Plant*, final report, Department of Trade and Industry, 2002; in PDF from http://www.dti.gov.uk/cct/cctdemohome.htm.

seem to have more fundamental attractions. Certainly the potential for energy savings from the introduction of more efficient systems at the point of use is very large. This is in part because, until recently, energy has been relatively cheap and energy efficiency has mostly been ignored. Given the increased level of concern about climate change, new policies have now emerged, aimed at improving the efficiency with which fuels are used, thereby reducing emissions. For example, the UK Cabinet Office's Energy Review, published in 2002, suggested that the efficiency of energy use in the domestic sector could and should be increased by 20% by 2010 and a further 20% by 2020.[13]

There has already been some success in making significant savings in other sectors in the UK, notably industry, where energy demand has been cut by nearly 40% over the last three decades, chiefly on the basis of the cost savings that could be made by investing in new, more efficient, technology. However, savings like this may not be possible to achieve in all sectors, with the transport sector being the most obviously problematic, given the seemingly inexorable rise in vehicle ownership and mileage around the world. Nevertheless, overall, the Carbon Trust, an agency set up with government support to promote a low carbon UK future, has estimated that overall energy conservation measures could perhaps achieve up to half the target of a 60% reduction in carbon emissions by 2050 proposed by the Royal Commission on Environmental Pollution.

Certainly, the enthusiasts for energy efficiency believe that spectacular savings can be achieved. For example, it has been claimed that a mixture of efficiency measures and demand management measures could possibly offer overall 'factor 4' improvements in energy use in most sectors, and possibly more.[14] However, while this may be possible in theory, in practice there may be problems. Once the cheap and quick energy saving options have been exploited, the opportunity for further energy savings may begin to reduce, while the costs could increase. Moreover, while there will hopefully be many technical and operational innovations than can improve efficiency, it is hard to see how efficiency improvements can be replicated continually, so as to keep pace with the projected 2% pa increase in basic global energy demand into the future.

To take a large scale example from the transport sector, although the world aircraft fleet has doubled its fuel efficiency over the past 30 years, global air traffic has quadrupled since 1970, from 350 billion passenger miles to 1500 billion passenger miles a year, and is forecast to more than double or even triple by 2050. To put the issue starkly, while typically fuel efficiency has risen by around 3% pa, in the year 2000 the use of aviation fuel in the UK rose by 10%. Aviation is currently responsible for around 3% of global carbon emissions (and around 10% of total greenhouse gas emissions) and as a result of increased demand, according to the IPCC, carbon emissions from this sector could increase by 478% between 1992 and 2050. Unless demand is somehow curbed, it would require major technical fixes and efficiency improvements to reduce this.

Similar challenges exist for most other types of consumer demand. Indeed, in some cases the situation can be worse, owing to the so-called 'rebound effect', which can undermine attempts to reduce emissions by improving energy effi-

---

[13] Performance and Innovation Unit, *Energy Review*, Cabinet Office, HMSO, London, 2002.
[14] E. von Weizsacker, A. Lovins and H. Lovins, *Factor Four*, Earthscan, London, 1994.

ciency.[15] To put it simply, the money that domestic consumers save by adopting energy conservation measures may be spent firstly on maintaining higher temperature levels, and then on other energy intensive goods and services, like foreign holidays by plane. The result can be that at least some of the initial energy and emissions savings may be cancelled out. The exact scale of this rebound effect is debated, with optimists suggesting that, since most products and services are less energy intensive than direct energy use, the rebound effect associated with extra spending on goods and services might only account for a 10% reduction in emission savings that would otherwise have been achieved.

However, overall, it does seem likely that, all other things being equal, if any commodity becomes cheaper then more of it used. The optimists sometimes argue that there may be saturation levels for demand in the affluent industrial countries, so that any new wealth will be spent on increasing efficiency of energy use rather than on energy use. However, so far, at the macro economic level, energy use shows no sign of decreasing, despite dramatic increases in energy efficiency. Moreover, as more people around the world join in the race to material affluence, demand seems certain to continue to increase.

This is not to suggest that energy conservation is not vital, since it is foolish to waste energy that has been produced at such a large potential environmental costs and, in general, increased efficiency should make for economic competitiveness. However, as pressure for an improved quality of life from an expanding world population grows, the increased energy demand could outpace the increase in savings that can be made, even given major commitments to improved energy efficiency. It may be that, at best, energy conservation *via* the adoption of energy efficiency measures and demand management will only be able to slow down the rate of increase in global carbon emissions. To actually reduce it requires a switch to non-fossil fuels. The options are nuclear power and renewable energy.

6 Nuclear Power

At present, the world obtains around 6% of its primary energy from nuclear power plants, but it has been suggested that this could be expanded as a response to climate change, since nuclear plants do not generate carbon dioxide, at least not directly. There is believed to be sufficient uranium for several hundred years at current use rates, although, if the nuclear programme is expanded dramatically, to try to respond to climate change, then reserves would diminish rapidly, leading to the use of more expensive lower grade ore or possibly, at even higher costs, uranium extracted from sea water. Alternatively, recourse could be made to fast breeder reactors, which in effect stretch the uranium resource, so that nuclear could conceivably play a role for some while.

For nuclear power to be able to play an expanded role, it would have to overcome the problems that have seen it fall from favour in the last few decades. This is not the place for an exhaustive review of the complex and often contentious issues surrounding the use of nuclear energy, which have been the subject of

[15] H. Herring, Is energy efficiency environmentally friendly?, *Energy Environ.*, 2000, **11**, 313–325.

several recent publications from varying perspectives, but in summary the key issues seem to be cost, safety and security.[16–18]

The cost issue was brought into sharp focus when an attempt was made to privatize the UK's nuclear plants in 1990. With nuclear generation costs, in the private sector context, of up to 6 p/kWh it was clearly unviable compared with gas/CCGT, at less than half of that figure. A non-fossil fuel levy was imposed to meet the extra costs, over fossil fuel generation, of the 20% or so of the UK's electricity that was then coming from nuclear plants. This levy put around 10% on consumers fuel bills, although a small part of the levy, initially 2%, did go to support some renewable projects. The levy initially raised around £1.2 billion pa for nuclear, before being phased out in 1998, by which time all but the old MAGNOX plants had passed into private ownership. The industry claims that new technology will soon be available (notably the Westinghouse AP 1000 redesign of the pressurized water reactor) which will be able to generate at competitive prices, possible 3 p/kWh, but so far this claim has yet to be tested. Similarly, the claim that the new so-called 'advanced passive' designs will be safer has yet to be confirmed.[19]

Clearly, safety is a major issue, not least since it is one reason for the relatively high costs of nuclear plants, which can cost up to three times as much to build as conventional plants, with the safety and control systems accounting for up to half the nuclear generation costs. One of the attractions of adopting passive fail-safe features like convection cooling is that there would be less need for expensive, complex and potentially unreliable emergency control systems.

However, it is unclear whether, even if the nuclear industry can develop cheaper, safer reactors, that this will necessarily reduce the level of public opposition to nuclear power. Overall, typically around 70% of the public in most countries is against nuclear power, with opposition mounting after each major nuclear accident. A recent UK opinion poll found that 68% of those asked felt that the UK should not build more nuclear plants in the next 10 years.[20]

Public opposition, coupled with poor relative economics, meant that, following the Three Mile Island accident in the USA in 1979, no new nuclear plants were ordered in the US and, following the Chernobyl disaster in the Ukraine in 1986, most of W. Europe has backed away from nuclear, with Germany being the most recent country to opt for a nuclear phase-out. Even France, the one-time mainstay of the European nuclear industry, imposed a moratorium on new projects, following the election of its red-green government in 1997. The UK has a policy of 'diminishing reliance' on nuclear. Given that currently no new plants are planned, that amounts to a phase-out.

[16] Royal Society/Royal Academy of Engineering, *Nuclear Energy: the Future Climate*, London, 1999.

[17] M. Grimston and P. Beck, *Civil Nuclear Energy: Fuel of the Future or Relic of the Past?*, Royal Institute of International Affairs, London, 2000.

[18] D. Elliott, The future of nuclear power, in *Energy Systems and Sustainability: Power for a Sustainable Future*, ed. G. Boyle, B. Everett and J. Ramage, Oxford University Press, Oxford, 2003.

[19] British Energy, submission to the PIU *Energy Review*, HMSO, London, 2002.

[20] British Market Research Bureau, *Public Opinion Survey* for the Royal Society for the Protection of Birds, 2001.

Nevertheless, the nuclear industry still has ambitions for expansion. Some countries in Asia are considering expanded nuclear programmes, notably Japan and China, and Russia still has a nuclear programme. Moreover, the USA is trying to relaunch its nuclear programme and the nuclear industry is keen to see the UK follow suit, on the basis of avoiding carbon emissions.

Assuming the cost and power plant safety issues can be dealt with, for these ambitions to be realized there still remains the unresolved problem of dealing with long-lived nuclear waste, which has to be kept isolated from the biosphere for millennia. Whatever happens to the UK nuclear progamme, around 500,000 tonnes of nuclear waste of various types will be produced in the UK over the next century, including the wastes arising from the process of plant decommissioning. Similar problems exist around the world. However, at present, although some countries are exploring possible sites, no repositories exist for the indefinite storage of long-lived high-level nuclear waste. Moreover, perhaps inevitably, there is usually strong local opposition to any proposed site. In this situation, many feel that it would be irresponsible to expand nuclear power. The environmentalists argue: why try to solve one global environmental problem, climate change, by creating another—yet more radioactive pollution?[21]

The final issue is the increasingly worrying and topical problem of nuclear weapons proliferation and diversion of weapons-making material from the plutonium extracted from spent fuel. The UK has around 70 tonnes in store, some of it of high grade. Given nuclear expertise, a few kilos of weapons-grade plutonium can be used to make a crude bomb. Equally worrying, following the events of September 11th 2001, is the possibility of terrorist attacks on nuclear plants or interim waste stores, potentially releasing wastes over a wide area should the containment be breached.[22]

It may be that new technology and new operating procedures could reduce the amount of waste produced. Certainly the abandonment of plutonium extraction by reprocessing spent fuel could reduce the amount of medium- and low-level wastes produced, and keep the plutonium in a form that would make it difficult for terrorists to use. At present, only the UK and France are continuing with reprocessing on a significant scale. Given that the UK and France, along with the USA, have closed their breeder programmes, the only use for the plutonium, other than in weapons, is for a new 'mixed oxide' fuel, with Japan currently being the main customer. However, given that there are ample and much cheaper supplies of uranium, the market for this new fuel is likely to be limited, and trading with it around the world presents potential risks of terrorist attacks.[23]

While the nuclear industry still seems keen to promote nuclear fission as part of the solution to climate change, there is also the longer-term option of nuclear fusion. This is still at the R&D stage and, despite around £20 billion having been

[21] Nuclear Information and Resource Service, CDM—a new nuclear subsidy?, *Climate Change and the CDM*, briefing note, Washington, DC, 2001.

[22] STOA, *Possible Toxic Effects from the Nuclear Reprocessing Plants at Sellafield (UK) and Cap de La Hague (France)*, WISE–Paris report for the European Parliament Scientific and Technological Options Assessment programme, Strasbourg, 2001.

[23] POST, *Mixed Oxide Nuclear Fuel*, briefing paper, Parliamentary Office of Science and Technology, London, 2000.

spent on fusion research world-wide so far, at best a commercial reactor is still many decades away. Fusion is sometimes portrayed as being a cleaner and safer option than fission. However, most of the likely approaches to fusion would involve a plasma of radioactive tritium running at 200 million degrees Kelvin, and the risk and impacts of accidental release could be significant. In addition, the neutron radiation in the fusion reactor core would produce radioactive materials which would have to be stored. Although they would have shorter half lives than the products from fission, there would still thus be a waste problem to deal with. The science is certainly exciting but, in terms of energy supply, fusion is a long shot, with unknown costs and some safety problems and, perhaps more importantly, it is unlikely to be available in time to help deal with the urgent problem of climate change.[24]

7 Renewables

Fortunately, we may not need to, in effect, create little suns on Earth with fusion reactors. The sun is a working fusion reactor that already supplies more energy than human beings could use, if technologies can be developed to tap it efficiently. The incoming solar energy drives the climate system, creates winds, waves and rain for river flows and sustains biomass growth. The use of these climate and weather-driven energy flows and sources would represent a rather elegant closing of a circle, since we are talking about using the climate and weather system to help us avoid climate change by substituting for the use of fossil fuels.

Not all the natural energy flows available are climate driven. In addition to the continually renewed solar sources, the gravitational pull of the moon on the sea results in tides: lunar power. Although not strictly a fully renewable resource, there is also the heat deep underground created by radioactive decay, a source of geothermal energy.

The total energy available from the various renewable energy sources is very large. At present the total energy generating capacity of the various energy conversion systems built by mankind amounts to around 14 TW. The continuous solar input is equivalent to 90 000 TW, of which about 1000 TW could, in principle, be captured for energy conversion to forms we can use.[25] Of course, there are also efficiency losses and land-use constraints to take into account, but even so, there should be sufficient energy from these sources to met our needs many times over.

Estimates of the practical contribution that might be made by the renewable sources in the years ahead inevitably depend on the assumption made about the level of financial support, and the scale of constraints on the conventional and nuclear options. The renewables energy systems involve new technologies seeking to win a place in a market dominated by the existing energy technologies, with oil and increasingly gas dominating the energy market. Energy scenarios

[24] D. Elliott, *Energy Society and Environment*, Routledge, London, 1996; for a useful overview see the Parliamentary Office of Science & Technology's *Nuclear Fusion Update*, POST note 120, 1998 (http://www.parliament.uk/post/home.htm).

[25] T. Jackson, Renewable energy: summary paper for the renewable series, *Energy Policy*, 1992, **20**, 861–883.

are widely used to describe possible paths ahead and the sustainable growth scenario produced by Shell International in 1995 has been very influential. It suggested that, by around 2060, renewables sources could be meeting about half the world's total energy needs.[26] Subsequent studies have suggested that, in principle, by 2100, renewables could perhaps be meeting over 80% of global energy needs, assuming that they were seen as a priority for environmental reasons.[27]

Inevitably, long-term projections like this are very speculative. What matters in reality is practical progress towards technologically and economically viable energy systems.

The renewable contribution started out from a relatively low level, basically traditional hydro and traditional biomass. Even so it is worth noting that around 17% of the world's energy comes from these two sources. Although funding levels have been relatively low compared with nuclear technology, progress on the so-called new renewables has been quite rapid since the 1970s, when interest in these sources first emerged as a consequence of the oil price crises. By 2000, the new renewables were supplying around 2% of world energy, or about 3% of its electricity. The contribution is expanding rapidly, stimulated by some quite demanding targets. For example, the European Union is aiming to have 12.5% of its electricity produced from renewable sources by 2010, with some member countries aiming for much higher targets. Denmark is aiming at 29%, Finland 21.7%, Portugal 21.5% and Austria 21.1%, these figures excluding the contribution from large hydro.[28]

8 Progress with Renewables

Some comprehensive and up-to-date reviews of technological progress have been published recently.[29,30] Consequently, what follows is an overview of key developments in the main fields.

Wind power has been one of the leaders, with an initial market growth rate of over 20% pa, accelerating up to 30% pa after 1997. By mid-2002 there was around 23 000 MW of wind-turbine capacity in operation, with Denmark, one of the pioneers, obtaining around 18% of its electricity from wind projects. In Germany the rate of installation has risen to around 1.5 GW pa, with nearly 9000 MW in place by 2002. The technology has developed rapidly from small, fixed-speed 100–200 kW machines to variable speed machines rated at 2 MW and more. Similarly, generation costs have fallen dramatically. For example, whereas in the UK in the early 1990s wind projects were receiving 11 p/kWh under the Non-Fossil Fuel levy, by 2000 wind projects were going ahead in some locations at around 2 p/kWh, without the need for subsidy.

[26] Shell, *The Evolution of the World's Energy System* 1860–2060, Shell International, London, 1995.
[27] World Energy Council, *New and Renewable Energy Resources: A Guide to the Future*, Kogan Page, London, 1994; World Energy Council, *Energy: Our Common World—What Will the Future Ask of Us?*, conclusions and recommendations, 16th WEC Congress, Tokyo, 1995.
[28] European Union, *Directive on Renewable Energy, Indicative Targets*, 2002.
[29] R. Gross, M. Leach and A. Bauen, Progress in renewable energy, *Environ. Int.*, 2003, in press.
[30] G. Boyle (ed.), *Renewable Energy: Power for a Sustainable Future*, 2nd edn., Oxford University Press, Oxford, 2003.

The next stage is to go offshore, where environmental constraints are lower and wind speeds generally higher and more consistent. That can partially compensate for the increased cost of offshore location and of delivering power back to land. By 2002 there was nearly 100 MW of offshore capacity in place in the EU. The Netherlands, Denmark and Sweden have been pioneers, and they are now being followed by Germany, the UK, Eire, Belgium and France. For example, Germany has a very ambitious programme, with a target offshore wind capacity of 10 GW by 2030, supplying around 25% of the country's electricity. Overall the EU's total ultimate offshore wind resource has been put at around 900 TWh pa. Currently, offshore wind projects generate power at around 4–6 p/kWh, but prices are falling as experience is gained and new technology emerges, with capital cost/kW installed falling by around 30% over the past decade. Evidently then, the wind power success story is likely to continue.[31,32]

While wind power has attracted most media attention, the use of biogas from biomass wastes has actually proved to be the most immediately economic option, with sewage gas and gas from landfill sites being amongst the cheapest renewable energy sources. Although also economically attractive, energy recovery *via* the combustion of municipal and domestic solid wastes has been less successful, mainly due to opposition by local residents concerned about the potential for emissions of dioxin from the combustion of plastics. In addition, most environmentalists do not see these wastes as genuinely renewable sources since they rely on the production of materials which ought, they feel, to be avoided where possible, and recycled where not.

Specially grown energy crops are, by contrast, usually seen as attractive options by most environmentalists, as long as the rate of use is matched by the rate of replanting to maintain rough overall carbon neutrality. Liquid fuels, derived from rape seeds or sunflowers, have niche markets for transport use in the form of biodiesel in some parts of continental Europe. The UK is pressing ahead with an energy crop programme based on the use of wood chips from short rotation coppices as a fuel for advanced gas turbine plants. Costs are still relatively high, with the 10 MW ARBRE scheme in Yorkshire expected to generate at 8 p/kWh.

Photovoltaic solar (PV) is even further away from commercial competitiveness, with commercial cells generating power at around 5–10 times the cost of conventional power. However, there are major efforts underway to reduce costs and, since solar PV is widely seen as having a very significant future, major programmes of PV installation have been supported around the world, particularly in the USA, Germany and Japan. For example, Germany has a 100 000 PV roof programme. The main improvements are expected to be in relation to the use of new materials with higher efficiencies and lower costs. Current commercial devices have energy conversion efficiencies up to 12–15%, but lab devices have been developed with efficiencies up to 25%, and some multi-junction devices have even higher efficiencies. In terms of costs, ease of manufacture is an obvious

[31] P. Gipe, *Windpower Comes of Age*, Wiley, Chichester, 1995.
[32] R. Y. Redlinger, P. D. Andersen and P. E. Morthorst, *Wind Energy in the 21st Century: Economics, Policy, Technology, and the Changing Electricity Industry*, Palgrave/UNEP, 2002.

issue and thin film amorphous silicon technology has proven to be attractive, despite its lower efficiency. There are also exciting prospects for some of the newer polymer cells.[33]

However, the economics is not just a matter of the device. Equally, new applications and new patterns of energy generation and use are opening up new markets for PV. PV is rarely seen as a way to compete directly with conventional grid-linked power plants, but rather as providing a way to generate energy directly at the point of use, thereby avoiding transmission losses over long distance. In many developing countries, off-grid PV can be the only option, apart from imported diesel or petrol, for power generation at the village level. There are around 2 billion people in the world who are without access to electricity and who are never likely to get access to it *via* grid links. Even in the industrialized countries it may be that houses and offices will increasingly be equipped with PV cells on the roof, meeting most of their light electrical load, with any excess power being exported to the grid, and any shortfall being made up by power imported from the grid. Several examples already exist in the UK, with 'net metered' power actually making a profit for the home owner. Clearly for the moment, installing a PV roof still involves a significant capital investment, with payback times put in the decades. However, having a building-integrated PV roof made up of the new generation of PV tiles avoids the cost of a conventional roof, which can reduce overall costs, and it is certainly interesting to have a roof that earns its keep. Moreover, as take-up in niche markets expands and volume production grows, so prices should drop, so that the use of PV could become widespread within a decade or so.[34]

The newer renewables include offshore wave and tidal current power. The UK's wave energy resource is quite significant; possibly 50% of UK electricity could be generated from this source and maybe more if generators can be located further out to sea where the largest resource is to be found. Research into wave energy started in the UK in the 1970s. However, after some initial trials, it was claimed that the cost of generation would be high and the government-funded programme was wound down. Some work nevertheless continued, both in the UK and elsewhere, and recently there has been something of a renaissance for wave development. In the UK, Wavegen were the pioneers, with a 500 kilowatt LIMPET shoreline device on Islay commissioned in 2000. They are currently going offshore with a new floating device, backed by a £1.67M DTI grant. Ocean Power Delivery are testing their Pelamis wave snake device in Scottish waters. Elsewhere there is a wide range of device types under test, including the Dutch Archimedes Wave Swing and the Danish Waveplane. Japan has the floating 'Mighty Whale' test platform and a novel wave-focusing device is being developed in Australia.[35]

Tidal current technology is a more recent development. At one stage there was interest in the idea of building large tidal barrages across suitable estuaries to

[33] M. Gratzel, Photoelectrochemical cells, *Nature*, 2001, **414**, 338–344.
[34] KPMG, *Solar Energy: from Perennial Promise to Competitive Alternative*, report for Greenpeace, London, 2000.
[35] House of Commons Select Committee on Science and Technology, *Wave and Tidal Energy*, session 2000–2001, seventh report, HMSO, London, 2001.

create a head of water by trapping tidal flows. The Severn estuary was one of the most promising sites, capable of housing a barrage with 8 GW of generating power. France built a small barrage on the Rance estuary in the 1970s. However, the capital cost of barrages is very high and such projects, with their long lead times, are unattractive in the contemporary private sector investment context. By contrast, the use of small free-standing turbines mounted on the sea bed in the tidal flows is seen as a better option. They would be modular and much less environmentally invasive than large expensive barrages. The total resource for barrages and tidal currents is about the same in the UK. If fully developed, both could generate around 20% of UK electricity.[36]

There are still some novel ideas for building circular tidal basins offshore to capture high tides for barrage-type operation, but in the main the future for tidal power in the UK looks like being linked to the use of tidal flows. Prototypes are currently under test. For example, Marine Current Turbines are testing a sub-merged 300 kW tidal turbine off the coast of Devon, with an EU grant of 1M euros matched by £1M from the DTI, and a novel sea-bed-mounted hydro-plane system, the Stingray, is being tested off the Shetlands, with a £1.1M DTI grant.[37]

Marine energy sources like wave and tidal are more predictable and less variable than wind power. Waves are in effect stored wind energy, and the tides are driven by the regular lunar cycle. Practically, they could fit in well with the other major offshore energy option, wind power. For example, there have been plans for combining offshore wave or tidal devices with wind turbines, although there are some trade-off issues, since the best tidal and wave sites are not always the best offshore wind sites. However, they all share the same major advantage that they are offshore, so there is far less visual intrusion and the environmental impacts on marine life and marine ecosystems seems likely to be minimal. Indeed, some might be positive, in terms of providing a habitat for some sea creatures.[38]

While there are plenty of sites around the UK coast for tidal current devices, wave energy is more concentrated, most of it being off the North Atlantic coast. That implies that it would have to be landed mostly in Scotland. However, the main power demand is in S.E. England. Given there are constraints on building large new grid links across the north of the country, there is interest in the idea of an undersea high-voltage DC power cable from Scotland to England and Wales, which might add 0.5 p/kWh to the price of the power.[39]

So far, most of the wave and tidal device teams are expecting their first prototypes to deliver power at around 7 p/kWh, which is far better than on-land wind achieved at an equivalent stage in its development. Most of the wave and

[36] Energy Technology Support Unit, *New and Renewable Energy: Prospects in the UK for the 21st Century*, supporting analysis for the DTI energy review, ETSU R-122, 1999.

[37] Government reply to Science and Technology Committee report on *Wave and Tidal Energy*, 2001, and the subsequent Parliamentary debate on Jan 10th 2002.

[38] DEWI, *North Sea Offshore Wind—a Powerhouse for Europe: Technical Possibilities and Ecological Considerations*, produced for Greenpeace International, Amsterdam, 2000.

[39] PB Power, *Electricity Network Limitations on Large-Scale Deployment of Wind Energy*, report for the Energy Technology Support Unit, ETSU W/33/00529/REP, 1999; PB Power, *West Coast Interconnector Study*, concept paper for ETSU, 2002.

tidal teams hope that commercial scale devices will deliver at around 4 p/kWh, making them viable for take-up under the Renewables Obligation (three of the wave projects already have contracts under the Non-Fossil Fuel Obligation).

Although much of the initial work has been done in Europe, the marine renewables are being deployed around the world. There are wave energy projects in India and the Azores and the Canadian company, Blue Energy, has an ambitious tidal fence planned for installation in a causeway between islands in the Philippines, which could ultimately be expanded to deliver 2200 MW. There have also been even more ambitious proposals, including the idea of using arrays of submerged turbines to tap the very large energy flows in the Gulf stream.[40]

There are many other renewable energy options on the large and small scale. While concerns have begun to be raised about the environmental impact of large-scale hydro projects, small-scale hydro is widely seen as a useful option. Although it is not strictly renewable, since the local heat flux can be exhausted at least for some while, there is around 7 GW of geothermal electricity generating plant installed around the world. In addition, geothermal sources are widely used for heating, as of course is direct solar energy.

Roof-top solar collectors for water heating are ubiquitous in many of the sunnier parts of the world, but can also play a role for space or water heating in other climates. Passive solar design has an even larger role to play. Solar heat can also be concentrated to power so-called 'solar thermal' electricity generation plants, several of which have been constructed. There is also the more novel idea of building solar chimneys, with the updraft created by a large ground level conservatory driving an air turbine mounted in the chimney stack. In addition, there are a variety of biogas and biofuel options; for example, anaerobic digestion of biological materials and pyrolitic conversion to liquid fuels, which can provide fuel both for heating and for power production and, in some cases, for use in vehicles. For example, it may be possible to generate hydrogen gas from biomass and wastes. While electricity production has tended to dominate the commercial development of renewables in recent years, there are clearly just as many options for production of heat and also of fuels for transport use.[41]

9   Assessing the Costs of Sustainable Energy

The forgoing review of technological options should have highlighted the large range of energy options that exists and, to the extent that they compete for funding, the need for some way to make choices between them. This is particularly difficult since they are all at different stages of development. As we have seen, most of the new renewables are still at a relatively early stage, and yet some at least are likely to have to play a major role in the future.

Economic assessment, in terms of the cost of energy produced, is the most obvious way to try to compare the viability of energy systems. However, making judgements as to which options to emphasize based on prototypes at the early

[40] D. Elliott, Tidal power, in *Renewable Energy: Power for a Sustainable Future*, ed. G. Boyle, Oxford University Press, Oxford, 2003.

[41] T. Johansson, H. Kelly, A. Reddy and R. Williams, *Renewable Energy: Sources for Fuels and Electricity*, Earthscan, London, 1993.

Table 1 Cost of electricity in the UK in 2020

Source	Pence/kWh
On-land wind	1.5–2.5
Offshore wind	2–3
Energy crops	2.5–4
Wave and tidal power	3–6
PV solar	10–16
Gas CCGT	2–2.3
Large CHP/cogeneration	under 2p
Micro CHP	2.3–3.5
Coal (IGCC)	3–3.5
Nuclear	3–4

Source: UK Government Cabinet Office.[13]

stages of development clearly has its limitations, as was demonstrated in the case of wave power in the UK. Early cost estimates of 20 p/kWh or even 50 p/kWh, at a time when only small models had been tested, led to the withdrawal of funding, first in 1982 for deep-sea wave and then in 1994 for the remainder of the programme. It was almost 20 years before a re-assessment suggested that much lower prices might be attained, with, in March 2001, the DTI commenting '*The decision was taken in the light of the best independent advice available. With the benefit of hindsight, that decision to end the programme was clearly a mistake*'.[42]

Of course, the validity of that reassessment is still to be demonstrated. However, in general, rather than trying to 'pick winners' at an early stage, it seems wiser to allow a range of developments to proceed, especially since the level of expenditure at the R&D stage is relatively small. Selection becomes more important at the next, more expensive stage in the innovation process, when moving on the full-scale development and commercial deployment. Some renewables have reached that point, and market assessments are therefore beginning to be made, but most are further back in the process.

Although current prices are sometimes a poor guide to what prices might be in the future, it is still possible to identify trends. The Energy Review carried out by the Cabinet Office's Performance and Innovation Unit made use of learning curve analysis to try to identify trends in price reductions. If prices at successive stages in the innovation process are plotted against cumulative production volume on a log-log scale, then in many cases a straight line results. The slope varies with the technology, but the range is not great: gradients of between 10 and 20% are typical. Not all the energy options reviewed could be assessed in this way, and in some cases parametric engineering assessments and proxy assessments had to be used.[13]

The results for some of the key new renewables, compared with conventional sources, are shown in Table 1. Clearly these long-term estimates are speculative and rely on a range of assumptions about policy developments. For example, if funding is not provided for new renewables or new nuclear technologies, then the picture could look very different.

[42] D. Ross, Scuppering the waves: how they tried to repel clean energy, *Sci. Public Policy*, 2002, **29**, 25–35.

Table 2 Extra cost resulting from environmental damage	Cost to be added to conventional electricity cost (assumed as 0.04 euro/kWh average across the EU) in euro/kWh	
	Coal	0.057
	Gas	0.016
	Biomass	0.016
	PV solar	0.006
	Hydro	0.004
	Nuclear	0.004
	Wind	0.001

Source: EU EXTERNE study.[43]

One of the main uncertainties is over how the relative environmental costs of the various options will be assessed in future. If, for example, the full social and environmental cost of carbon dioxide and acid emissions from coal combustion is added to the cost of generation, then the comparisons would alter dramatically. As we have seen, although nuclear plants do not generate carbon dioxide there are other impacts and risks with the nuclear option. Moreover, even the most benign renewable sources will have local impacts.

To help make comparisons, the EU EXTERNE study has tried to put a price on the environmental and social impacts of energy use, focusing on electricity production in the EU. The results suggest that the cost of producing electricity from coal would double and the cost of electricity production from gas would increase by 30% if external costs such as damage to the environment and to health were taken into account. It is estimated that these costs amount to around 1–2% of the EU's Gross Domestic Product (GDP), not including the cost of damage caused by global warming.[43]

The methodology used to calculate the external cost is called impact pathway methodology. This methodology sets out by measuring emissions (including applying uniform measuring methods to allow comparison), then the dispersion of pollutants in the environment and the subsequent increase in ambient concentrations is monitored. After that, impact on issues such as crop yield or health is evaluated. The methodology finishes with an assessment of the resulting cost.

The EXTERNE methodology could be applied to other energy-related sectors like transport. In fact, preliminary work has shown that aggregated costs of road transport, the dominant source of damage, add another 1–2% of GDP to the bill.

The results for electricity are shown in Table 2. As can be seen, on the basis of this analysis, fossil fuels have much larger environmental impacts than any of the other options, with coal clearly being the worst. By contrast, wind is relatively benign, having four times lower environmental costs than nuclear.

Assessments like this are fraught with methodological problems, for example, concerning how to calculate the cost of specific types of damage. Simple economic assessment, based on insurance replacement costs, may not provide a realistic measure of, or proxies for, the human value of amenity loss or health damage, much less the ecological value of any disruption. Even more contentious

[43] European Commission, *Externalities of Energy*, reports on the EXTERNE programme, DG12, L-2920, Luxembourg, 2001.

Table 3 Comparative risk of electricity production by fuel cycle

Number of deaths and diseases per GWyr, including entire fuel cycle but excluding severe accidents

	Occupational hazards		Public (off-site) hazards	
	Fatal	*Non-fatal*	*Fatal*	*Non-fatal*
Coal	0.2–4.3	63	2.1–7.0	2018
Oil	0.2–1.4	30	2.0–6.1	2000
Gas	0.1–1.0	15	0.2–0.4	15
Nuclear (LWR[a])	0.1–0.9	15	0.006–0.2	16

[a]LWR = Light Water Reactor, the generic term for reactors using ordinary water for cooling, like PWRs.
Source: W. Nordhaus.[48]

is the value put on human life, which, inequitably but seemingly inexorably, differs around the world.

Clearly then, some of these valuations are subjective and this is a problem which has perhaps even more influence on popular assessments of the impacts of renewable energy technologies. For example, although the scale of the local impacts from most renewables is small compared to the massive global impacts of burning fossil fuels, they can nevertheless lead to public concerns, as has happened in the case of local negative reactions to visual intrusion by some wind power projects in the UK.[44]

10   Risks

The problem of making impartial judgements of impacts and costs can be even more complex in relation to assessing the impact of accidents and health effects and their costs, one of the subset of assessments made within the EXTERNE analysis. The impacts are clear enough in general terms. Nuclear power introduces novel health risks related to the generation of long-lived radioactive materials, while there are large public health implications from the emission of acid gases from fossil fuel combustion. In addition, there are safety risks with some renewables, such as large hydro.

Unfortunately, however, the scale of the health impacts and the degree of risk are often contentious, with disagreements on the data and its analysis being common, as witness the long debate over the results of early attempts to assess the risk of nuclear plants[45] and energy sources generally.[46,47]

[44] D. Elliott, Public reactions to wind farms: the dynamics of opinion formation, *Energy Environ.*, 1994, **5**, 343–362.

[45] N. Rasmussen, *The Safety of Nuclear Power Reactors (Light Water Cooled) and Related Facilities*, WASH-1250, 1973; *An Assessment of Accident Risks in US Commercial Nuclear Power Plants*, WASH-1400, 1974; *Reactor Safety Study: An Assessment of Accident Risks in U.S. Commercial Nuclear Power Plants*, WASH 400/NUREG 75/014, Washington, DC, 1975.

[46] H. Inhaber, *Risks of Energy Production*, Canadian Atomic Energy Control Board, AECB-1119, 1978; *Energy Risk Assessment*, Gordon & Breach, New York, 1982.

[47] J. P. Holdren, *Risk of Renewable Energy Sources: a Critique of the Inhaber Report*, University of California, report ERG 79-3, 1979.

Prof. William Nordhaus, a US based economist, and a proponent of nuclear power, has produced the analysis of the death and disease rates associated with the use of fossil and nuclear fuels as set out in Table 3.[48]

The most striking thing is the very large number of off-site public injuries and deaths/GWyr associated with coal and oil, compared with nuclear. Gas also comes out quite well on these comparisons. These figures reflect the significant impacts on public health (for example, in terms of respiratory diseases) of the acid and other emissions from coal and oil burning, and the lower impacts of gas. In terms of occupational hazards, it is worth remembering that some of the nuclear occupational deaths and illnesses covered by Nordhaus will be associated with uranium mining, which have to be set alongside the deaths and injuries from coal, gas and oil extraction activities.

However, the Nordhaus analysis excludes 'severe accidents', which, given Chernobyl, might be thought to undermine its credibility. Unfortunately, there is some debate over the ultimate death toll from that accident. 31 People died immediately, or shortly after the accident, but initial estimates suggested that there could be up to 40 000 early (*i.e.* premature) deaths in the years ahead.[49] Certainly, subsequently, there have been reports of serious health impacts and continuing deaths attributed to the accident, for example amongst the workforce of approximately 200 000 'liquidators' who were hired or ordered to gather and bury radioactive material released by the blast. A study by the International Atomic Energy Agency, the World Health Organization and the European Commission, produced on the tenth anniversary of the accident, projected that this population of workers would suffer an excess burden of 2500 cancers as a result of their clean-up work. Moreover, it suggested that the residents of communities off-site might also be expected to suffer an excess burden of 2500 cancers as a result of exposure to fallout from the accident.[50]

However, some more recent reviews have produced much lower estimates. For example, a study by the UN Scientific Committee on the Effects of Atomic Radiation published in 2000 concluded that, apart from the initial deaths and 1800 cases of thyroid cancer in children, most of which were seen as potentially treatable, there was '*no evidence of a major public health impact*'.[51] It was suggested in a subsequent UN report that some of the illnesses emerged as the result of the stress of over-zealous evacuation and forced relocation, and that some could even be put down to psychosomatic effects.[52] Clearly there is room for disagreement about these assessments.

What about renewable sources of energy? Surely there is less room for debate there. One estimate of fatalities associated with energy production, reported by the Uranium Information Centre, puts the figure for accidents (such as dam

---

[48] W. Nordhaus, *The Swedish Nuclear Dilemma: Energy and the Environment*, Resources for the Future, Washington, DC, 1997.

[49] Z. Medvedev, *The Legacy of Chernobyl*, Blackwell, Oxford, 1990.

[50] International Atomic Energy Agency, the World Health Organization and the European Commission conference, *One Decade After Chernobyl*, 1996.

[51] UNSCEAR report to the General Assembly by the UN Scientific Committee on the *Effects of Atomic Radiation*, New York, 2000.

[52] UNDP/UNICEF, *The Human Consequences of the Chernobyl Nuclear Accident*, United Nations Development Programme/UN Children's Fund, 2002.

failures) associated with large hydro plants at 4000 immediate deaths amongst the public world-wide during 1970–1992.[53]

Given that the total amount of electricity supplied by hydro and nuclear plants world-wide is roughly the same, it is perhaps reasonable to compare the hydro accident figure with the deaths associated with the Chernobyl accident. However, in addition to these figures for major accidents for nuclear and hydro, it should be noted that the normal operation of the nuclear fuel cycle will lead to nuclear related deaths and injuries, as reflected in Nordhaus data above, which will not occur with hydro plant operations.

Comparisons are harder to make with wind power, since this is a relatively new technology. So far, with, in early 2002, around 23 GW installed (compared with around 350 GW of nuclear capacity), there have been around 21 operator deaths associated with wind turbines around the world, mostly due to falls and injury by blades, and no off-site injuries to the public.[54]

Clearly, in all but the most straightforward cases, making comparisons of risks is fraught with statistical and conceptual problems, and disagreements abound over the interpretation of data. What initially might seem like a simple, if grim, exercise in body counting turns out to be a far more complex and conflict-laden activity. We are therefore still some way from having a reliable approach to making comparisons of impacts amongst the options.

11 Carbon Accounting and Energy Analysis

Rather than trying to assign costs to impacts or measure health risks, another increasingly popular approach in attempting to reflect the environmental significance of energy technology is to use the resultant carbon emissions. If nothing else, this might help reduce the level of subjectivity involved in making comparisons between the various energy options. In effect we are thus moving one step backwards in the EXTERNE analysis.

Certainly, carbon emissions are a central factor in climate change, and it could be argued they can be used as a proxy for most other types of impact. One early estimate put the total fuel cycle emissions for coal-fired plant at 1058 tonnes of carbon dioxide per gigawatt hour, and 824 tonnes for combined cycle gas fired plants, compared with nuclear at 8.6 tonnes, wind at 7.4 tonnes and 6.6 tonnes for hydro.[55] A more recent lifecycle study by Hydro Quebec, published in 2000, covers all greenhouse gas emissions and translates them into equivalent carbon dioxide terms. It puts emissions at 974 tonnes/GWh for coal plants and 511 tonnes/GWh for combined cycle gas plants, compared with 15 tonnes/GWh for both nuclear plants and hydro dams and 9 tonnes/GWh for wind plants.[56]

A study on greenhouse gas emissions by the International Atomic Energy

---

[53] Uranium Information Centre, *Nuclear Electricity*, 6th edn., Melbourne, 2000.

[54] P. Gipe, Wind related deaths data base, *WindStats Newsletter*, 2001, **14**, no. 4.

[55] Meridian Corp., *Energy System Emissions and Material Requirements*, prepared for the Deputy Assistant Secretary for Renewable Energy, US Department of Energy, Washington, DC, 1998.

[56] Hydro Quebec, *Comparing Environmental Impacts of Power Generation Options*, greenhouse gas emissions fact sheet, 2000; PDF from http://www.hydroquebec.com/environment.

Agency in 2000 summarizes the ranges as follows: gas plant, 439–688 grams of carbon dioxide equivalent/kWh; coal plants, 966–1306 g/kWh; nuclear plants, 9–21 g/kWh; and wind plants, 10–49 g/kWh.[57]

To be comprehensive, carbon accounting must include not just the emissions during operation, but also the emissions associated with the energy use in construction of the plant and the materials used in its construction, as well as the energy used in decommissioning and (where relevant) waste disposal. Full 'life cycle' energy analysis of this sort is becoming increasingly important for all products and systems as part of their environmental assessment. In the case of power plants, the results can be revealing. For example, the Hydro Quebec study mentioned earlier suggested that the overall energy output to input ratio for nuclear was only 16, compared to 39 for wind. The figure for coal plants was 11, gas/CCGT 14, while that for hydro dams was put at 205, presumably because of the long lifetimes and large outputs of the plant. PV solar and biomass plantations had the worst ratios at 9 and 5, respectively, reflecting the large energy debt incurred in PV cell manufacture and the mechanical energy used with the harvesting and transportation of energy crops.[58]

Carbon accounting is becoming increasingly popular, given the various proposals for carbon emission permits and carbon trading arrangements that emerged from the Kyoto Climate Change Accord.[59] However, while comparisons of the carbon and greenhouse gas emissions are useful, and central in terms of global climate change, there are also clearly other impacts to consider in order to give a complete picture, not least acid emissions and radioactive emissions. There is thus a danger that seeking to optimize for low carbon emissions may in fact be sub-optimal for the environment as a whole. Of course, it could reasonably be argued that climate change is so important that other issues must take second place. However, even so, in terms of assessing the various low carbon options, the other impacts become significant. This is clearly the case in terms of radioactive pollution from the nuclear fuel cycle, although unfortunately that takes us back to the health issues discussed in the previous section and to debates about the impact of low-level radiation.

By contrast, there seems to be more opportunity for clear analysis in relation to renewables. Certainly, most renewables have low carbon emissions, so that local ecosystem effects (*e.g.* in relation to disruption of wildlife, biodiversity and, in the case of water flows, erosion) could be more important in choosing amongst them. A parametric approach to making comparisons of direct impacts has been developed in the case of renewables, based on the degree of local disruption of the natural energy flux that is involved.[60] On this measure, solar devices have very low impacts and at the other end of the scale hydro dams and tidal barrages have large impacts. In part this is because the latter two attempt to extract a large

---

[57] International Atomic Energy Agency, as reported on the Word Nuclear Associations web site at: http://www.world-nuclear.com/education/ueg.htm, 2000.

[58] Hydro Quebec, energy pay back fact sheet, 2000; PDF from http://www.hydroquebec. com/environment.

[59] UN Framework Convention on Climate Change, Kyoto Protocol, agreed at the third conference of parties held in Kyoto, Japan, 1997.

[60] A. Clarke, Comparing the impacts of renewables, *Int. J. Ambient Energy*, 1994, **15**, 59–72.

proportion of the very concentrated natural energy flows, whereas solar only intercepts a small proportion of a diffuse flow. Wave and wind devices fall in between these extremes, in terms both of energy extracted and the scale of the impacts. Wave devices attempt to abstract quite large amounts of the high flux incident energy and can have moderate impacts, but wind devices only extract a relatively small part of a low-density energy flow and have low impacts. The key issue, on the basis of this analysis, is to look at what the energy flows normally do in the local ecosystem, and then assess how much of this energy can be extracted without unduly disturbing key natural processes.

Hydro has a special problem in that it seems that in some locations (for example, in warm climates) the anaerobic digestion of biomass, brought continually downstream and tapped by the dam, can create methane gas to such an extent that a coal-fired plant of the same capacity would produce less net greenhouse gas impact.[61] What we are seeing here is the result of the disruption of a natural energy flow, which previously ensured the continual agitation of the water, so that anaerobic processes were minimized.

It is relatively easy to see how this energy flow functional analysis approach can be applied to 'flow'-type renewables like wind, hydro and wave power, but it may also apply to those involving 'stocks' of renewable energy, such as biomass. The issue then becomes the ecological value of the material being used, rather than just its energy value. Although it can be carbon neutral if the rate of burning is balanced by the rate of planting, the use of biomass has potential impacts, both from emissions produced by its combustion and because combustion destroys valuable organic material. These problems are shared by combustion of solid domestic and municipal wastes. Indeed, given the presence of plastics in the waste, the toxic emission problem can be much worse. However, on the basis of this functional analysis, the main problem could be the sterilization of valuable organic material.[62]

It could be that this functional analysis can also be applied to conventional fuels, albeit at a rather general level. For example, it could be argued that whereas when the energy in fossil and nuclear fuels was safely trapped underground it had no environmental impact, once released in the form of heat, combustion products and/or radionuclides, its pathway through the ecosystem involves environmental risks.

12 Land Use

Another, perhaps more concrete, way to assess the merits of energy systems on a quantitative basis is to compare their land usage. Given that there are competing uses for land, including obviously food production, but also increasingly housing, industry and leisure, this criterion could become more important. Since most renewable energy flows are diffuse, the collection technologies are likely to take up more room than conventional energy technologies.

---

[61] World Commission on Dams, *Dams and Development: A New Framework for Decision-making*, London, 2001.

[62] A. Clarke and D. Elliott, An assessment of biomass as an energy source: the case of energy from waste, *Energy Environ.*, 2002, **13**, 27–55.

However, there is a wide range of land use implications.[63] Energy crops (*e.g.* short rotation coppicing of willow) are the most land-using renewable source. Depending on location, coppices can take up to 20 times more land per kWh eventually generated than wind farms, and this may matter if land is scarce. Although it is currently still expensive, PV solar (on roof tops, so there is actually no real land-use implication) can also deliver more energy/hectare than energy crops, which is not surprising when the energy conversion efficiency of PV (up to 15%) is compared with the efficiency of photosynthesis (less than 1%). Indeed, according to the Hydro Quebec review mentioned earlier, PV comes out even better than wind, by a factor of nearly 2 in terms of energy/hectare.[64] Although energy crops can be stored, which gives them an advantage over intermittent sources like wind and solar, they are a bulky fuel, which will usually have to be transported to combustion plants, which can have implications for local road infrastructures.

Hydro dams have obvious land-use implications (since areas are flooded to make reservoirs), a problem not shared by tidal barrages. However, the latter can have significant impacts on the local and even regional ecosystem, so some land-use changes might occur, although some impacts may actually be positive. Offshore wind, wave and tidal current devices have no land-use implications, although, if they are near shore, there can be visual intrusion issues.[65]

While the visual impacts of the offshore options are likely to be very low, wind farms on land are seen by some as ugly and intrusive, and energy crop plantations may also prove to be unpopular if they cover large areas. However, these are human perceptual and aesthetic judgements, possibly also reflecting instrumental concerns (*e.g.* the belief that house prices will fall).

Here we are returning to subjective, perceptual concerns, and to debates over values and, in the end, personal and political priorities. It could be argued that if people want more energy they must be prepared to accept some form of intrusion and that, for example, wind farms are one of the most environmentally benign options (as well as being economically attractive). This argument may, however, not be accepted by those who see any disruption to treasured views as deplorable, and 'anywhere but here' as a viable policy for sustainable energy. Nevertheless, there is clearly a problem of aesthetics to address and a need for careful location and sensitive consultation.[66]

Opposition to wind projects has been particularly strong in some parts of the UK. However, the scale of this opposition has to be put in perspective. Regular opinion surveys have indicated overwhelming support for wind energy in principle, typically 70–80% being in favour. Indeed, the most recent survey, carried out in 2001 for the Royal Society for the Protection of Birds, found that only 3% of those asked were opposed to building wind farms on land.[20] However, campaigns by local opponents have been very effective at slowing down the wind

[63] D. Elliott, Land use and environmental productivity, *Renew*, Sept–Oct 2001, 22.

[64] Hydro Quebec land use fact sheet, 2000; PDF from http://www.hydroquebec.com/environment.

[65] A. Clarke, *Environmental Impacts of Renewable Energy: a Literature Review*, OU Technology Policy Group report, 1995.

[66] M. Pasqualetti, P. Gipe and R. Righter (eds.), *Wind Power in View: Energy Landscapes in a Crowded World*, Academic Press/Elsevier, London, 2000.

programme. Around 70% of project proposals have been turned down in recent years. It is interesting, by comparison, that in Denmark, where around 80% of the wind projects are locally owned, some by local co-ops, the level of opposition has been much lower. Similarly in Germany, most projects are locally owned and, as noted earlier, Germany has installed nearly 9000 MW compared with just over 500 MW in the UK, where there is only so far one wind co-op. Evidently, having an economic stake in the projects changes attitudes.[67]

To be fair, some opponents of wind projects are genuinely concerned about nature conservation and some believe that there are better options, including energy conservation or perhaps offshore wind. However, if the scale of the climate change problem is as large as some predict, and if we wish to avoid facing the risks of nuclear power, then there will be a need for all the renewable energy sources as well as all the energy savings that can reasonably be mustered. It may be that, as already noted, carbon sequestration can also provide some help, by storing some of the emissions produced by the continued use of fossil fuels. However, in terms of new energy supply technologies for the longer term, renewable energy seems to be the most promising option. In which case we may have to get used to landscapes in which windmills are, once again, common sights.

13 Choices for a Sustainable Energy Future

All technologies have impacts, and that implies that there is a need to make choices. In this review we have seen that energy technologies can be ranked on the basis of carbon dioxide emissions, which some see as not only the crucial environmental issue, but also as a useful proxy for other environmental impacts. Nuclear and renewables come out well in this comparison. However, this type of comparison leaves out the potential for major accidents and associated health risks. It is hard to see how many of the renewables, large hydro apart, could impose risks on the general public on the same scale as do nuclear power plants. Nevertheless, nuclear plants do not generate carbon dioxide, so there remains some interest in this option, although the issue of what to do with the radioactive wastes that are produced has yet to be resolved. Certainly most environmentalists argue that it would be foolish to try to deal with one problem (climate change) by creating another (radioactive pollution).

Carbon sequestration may offer a way to store emissions for a while, but underground storage may be not a reliable option for the long term. While much can be achieved by using energy more efficiently, we will still need to generate energy, and in terms of impacts, renewable energy seems like the best supply option for the future.

In the Energy Review carried out by the UK Government's Cabinet Office, the Performance and Innovation Unit concluded that the *'immediate priorities should be on energy efficiency and promoting renewables'*, although it added that the clean coal/carbon sequestration and nuclear options should be kept open in case renewables or energy efficiency did not deliver as much as was hoped.[13] A

---

[67] D. Elliott and D. Toke, A fresh start for windpower?, *Int. J. Ambient Energy*, 2000, **21**, 67–76.

similar policy was also supported by the major 'World Energy Assessment' carried out by the UN Development Programme, the UN Department of Economic and Social Affairs and the World Energy Council. It concluded that '*if the energy innovation effort in the near term emphasizes improved energy efficiency, renewables, and the decarbonized fossil energy strategies, the world community should know by 2020 or before much better than now if nuclear power will be needed on a large scale to meet sustainable energy goals*'.[7]

Assuming this approach is adopted, there then emerges a set of strategic issues concerning how best to develop renewables, which options to develop and on what scale. Most of the renewables are relatively small scale, compared with the technologies which have gone before, such as 1.2 GW nuclear and coal plants. However, as a result of the dash for gas, the trend has been to smaller multi-megawatt scaled plants, and some renewables are now operating at this level. For example, typically wind farms have around 10–20 MW of generating capacity and the ARBRE biomass plant is rated at 10 MW. Some renewables are better exploited on a smaller scale, right down to the house or building level, notably roof-top PV solar and solar heat collectors.

It could be that we will see a shift to a more decentralized and dispersed energy system, using small-scale local 'micropower' generators, both in the developing countries and in the developed countries. In the latter, some power will continue to be generated by medium- to large-scale power stations remote from consumers (including, increasingly, offshore wind, wave and tidal plants), but some power will be generated by users direct, possibly feeding any excess into local area grid networks. The national power grid could then become more of a two-way networking system, balancing out local power generation and local power demands around the country. With some generating capacity being embedded in local grids in this way, and with some local needs being met from local sources, transmission losses over long distances can be reduced. In addition, linking up local sources in this way can help deal with the problem of the intermittency of some renewable sources: the grid, in effect, balances out local variations.[68]

At the level of renewable energy contribution existing at present in the UK, intermittent inputs present few problems for the national grid. The variations are hardly detectable by the grid operators, and are much smaller than the variations from the conventional inputs, and the variations in demand. However, as and when the renewable contribution grows beyond around 20–30% of total electricity demand, then intermittency becomes more of a problem. Energy can be stored in batteries, flywheels, as compressed air or by pumped storage of water, as at Dinorwic, but energy storage is at present expensive. It may be that an interim option would be to use cheap, small, fast start-up gas turbines as back up, possibly fuelled by biomass or biogas. Another option is fast start-up fuel cell devices, or the novel Regenesys energy storage device.

However, as and when we move over to using hydrogen as a new fuel, then this could provide a major storage medium for power from intermittent renewable sources, with hydrogen being generated by the electrolysis of water. As long as

---

[68] C. Hewitt, *Power to the People*, Institute for Public Policy Research, London, 2002.

basic safety procedures are followed, hydrogen has many attractions as a new energy vector. In addition to being storable, it can be transmitted along gas grids, possibly mixed in with the remaining methane, at low cost and with low energy losses. It can be burnt directly as a heating or transport fuel, or to generate electricity in a power plant, or used to power a fuel cell, for houses or for vehicles. Hydrogen could thus become the key to a sustainable energy future, linking up with a range of renewable energy technologies.

A major attraction of the renewable is that there are so many options on so many different scales. The risk of failure is thus spread across a range of differing areas. Most of the plants can be installed quickly on a flexible, modular basis, and most can be easily decommissioned or removed if necessary, with minimal environmental disruption and no wastes. These attributes were evidently seen as important by the PIU, who commented: '*the desire for flexibility points to a preference for supporting a range of possibilities, rather than a large and relatively inflexible programme of investment such as is being proposed by the nuclear industry*'.[13]

Of course, there may be new developments which change this assessment at some point in the future. For example, there has been some interest in the development of smaller nuclear plants, like the 110 MW Pebble Bed Modular Reactor, being developed in South Africa with the support of BNFL and, until it recently withdrew from the project, the major US utility Exelon. It is claimed that the PBMR could be installed quickly and operate nearer to loads, thus making it suited to use in developing countries. However, it will be some years before this technology has been proven and, to the extent that nuclear power is adopted, or in the case of the USA re-adopted, it seems more likely that upgraded versions of conventionally sized plants will be used, like the 1000 MW Westinghouse AP 1000.

Clearly the mix of energy options used will depend on the local context and local resources. For example, for the UK, in its Energy Review the Cabinet Office's Performance and Innovation Unit suggested a target of obtaining 20% of UK electricity from renewables by 2020, with on-land wind and biomass being the main early entrants, possibly followed by PV solar and the marine renewables—offshore wind, wave and tidal current turbines. That would obviously make sense for a maritime nation with limited land area but excellent offshore energy resouces and well-developed offshore energy expertise.

By contrast, countries like China have very large land-based renewable resources, so a different pattern of development might be expected. Given its population size and rapidly growing economy, China is probably the pivotal country in terms of energy patterns in the industrializing world. China's on-land renewable energy resource, leaving aside large conventional hydro, is put at the equivalent of over 400 GW, which is more than the currently installed conventional generating capacity. The new renewable options include over 90 GW of small hydropower, about 250 GW of wind, approximately 125 GW of biomass energy, about 6.7 GW of geothermal energy and an abundance of solar installations. The current contribution from new renewables is around 19 GW, with most of this from small hydro, accounting for around 5% of total electricity. Large hydro supplies around 18%. For the future, wind looks like being the biggest growth area for China. On current plans, wind is expected to expand from 500 MW at present to 3 GW by 2005 and 5 GW by 2010. Small hydro is expected to rise to 22 GW by 2005 and 25 GW by 2010. By 2005, the total renewable capacity could be around 26 GW, rising to over 30 GW by 2010.[69]

Although China's economic modernization and rationalization programme has improved the efficiency of its energy and industrial systems, demand for energy is increasing rapidly. At present, the bulk of China's energy is produced from coal and there are large reserves. However, the combustion of this coal is already creating major air quality and health problems, as well as contributing to carbon dioxide emissions. While, as noted above, there are very large renewable resources and a programme of expansion, China is also keen to expand its nuclear programme. Whether it can or should expand both remains unclear.

Whereas it seems that most of Europe has made its choice, and is pushing ahead with renewables rather than nuclear, China, like may other rapidly industrializing countries around the world, is, it seems, still at an energy cross-roads.

14   Conclusions

Although there are clearly many choices to be made, and many difficult technical and economic problems to overcome, there is nevertheless something of an emerging consensus that, as the UN/World Energy Council 'World Energy Assessment' report, published in 2000, put it: *'there are no fundamental techno-logical, economic or resource limits constraining the world from enjoying the benefits of both high levels of energy services and a better environment'*, although, a little more cautiously, it added: *'A prosperous, equitable and environmentally sustainable world is within our reach, but only if governments adopt new policies to encourage the delivery of energy services in a cleaner and more efficient way'.*[7]

The threat of climate change has evidently galvanized most of the world's leaders into making commitments to reduce emissions, with most industrialized countries, the US notably apart, supporting the Kyoto Protocol, which calls for a reduction in greenhouse gas emission by around 5% compared with 1990 levels, to be achieved during the period 2008–2012. Within this framework, some countries are opting for larger reductions. The UK, for example, has volunteered to cut carbon dioxide emissions by 20% by 2010. What remains unclear is whether these targets can be reached, what further reductions will be seen as necessary and viable in the future, and what role will be played by the various energy options discussed in this article.

As we have seen, the UK government is considering a target of obtaining 20% of its electricity from renewables by 2020. Most of the rest of Europe could probably do even better than that. The EU Renewables Directive suggests that some should achieve 20% or more, leaving aside large hydro, by 2010. Interest-ingly, Scotland is already considering a 30% target for 2020 and it has been suggested that the UK as a whole could also achieve this target.[70]

[69] L. Hongpeng, *Renewable Energy Development Strategy and Market Potential in China*, World Renewable Energy Congress VI, congress papers, Pergamon Press, Oxford, 2000, pp. 90–96.

[70] P. Ekins, *The UK's Transition to a Low Carbon Economy*, Forum for the Future, 2001.

Gas will, of course, remain the dominant fuel for some while, but nuclear seems likely to decline, at least in Europe. That may, of course, present interim problems in terms of carbon emissions, if renewables cannot be expanded fast enough to take over. However, there could be some interim solutions. For example, although, in the UK, around 9 GW of old nuclear plant is set to be retired over the next two decades, as noted earlier, by 2010 the UK is planning to have 10 GW of CHP capacity installed, with presumably more to follow. The widespread adoption of gas-fired CHP would provide heat (which would otherwise not have been available for use), which would release gas currently used for heating, for use in electricity generation. This could in effect replace the output of nuclear power plants as they retire, without leading to any increase in carbon emissions, leaving renewables to begin to replace coal burning and also to begin to provide electricity, hydrogen and biofuels for transport use. Carbon sequestration might be seen as performing a similar interim role, by allowing gas to replace nuclear without creating any extra emissions. In parallel, investment in energy efficiency and demand-side management should be able to hold down the rate of increase in demand, with, ideally, the cash savings from energy conservation being used to fund the expansion of renewables.

Clearly, the rapid expansion of renewables will be a major technological challenge, but equally it could be seen as a major opportunity for economic as well as environmental benefits, given the prospect of a large international market for renewable energy technology. It may not be entirely rhetorical to talk, as have some government ministers, of the UK being on the threshold of a 'green industrial revolution'.

Renewable Energy: Technology Considerations and Electricity Integration

DAVID INFIELD AND PAUL ROWLEY

1 Introduction

It is widely accepted that renewable energy (RE) sources are the key to a sustainable energy supply infrastructure since they are both inexhaustible and non-polluting. A number of RE technologies are now commercially available, the most notable being wind power, photovoltaics (PV), solar thermal systems, biomass and the various forms of water power. The wind power and PV sectors are currently experiencing significant investment and growth; they were the fastest expanding forms of electricity generation world-wide in the year 2001, with an overall sectoral expansion of around 40%. This anticipated growth is projected to increase significantly in the short term and raises important questions regarding the way in which these new sources can be used to supply society's energy needs. A key challenge is to integrate renewable energy generation into our electricity supply system without compromising safety or reliability.

This review examines the current state of renewable energy technologies, with particular emphasis on the work being done to ensure that they can be successfully integrated into electricity systems.

2 A Brief Introduction to Integration Issues

Aggregate Supply and Demand Issues

Renewable energy supplies are in general intermittent and, depending on the technology, are to some extent unpredictable. Wind energy is one of the most variable and most difficult to forecast, although there are benefits related to geographical diversity, as will be elaborated in Section 3. At the other end of the spectrum is biomass, which (despite an annual growth cycle for most energy crops) is entirely predictable, provided a sufficient store of feedstock is main-

Issues in Environmental Science and Technology, No. 19
Sustainability and Environmental Impact of Renewable Energy Sources
© The Royal Society of Chemistry, 2003

tained. In this it closely resembles conventional fuels (which can be burnt as required to meet demand), especially if primary biomass feedstock is used to produce secondary fuels, such as gasification or pyrolysis products. However, the intermittent renewables are unlikely to be controllable (dispatchable) in the way conventional power stations are. All of this means that the way in which such generators are integrated into the electricity supply infrastructure differs from accepted practice, and could, without proper understanding and planning, cause disruption to the electricity supply system.

It is useful to classify the impacts of renewable forms of generation into those that are system-wide (such as effects on system frequency and the related issues of conventional plant control, and dispatch), and local (such as voltage rise and variation effects). Generally these local effects will be experienced within the distribution, rather than the transmission system;* they will be discussed in the next sub-section.

Many people believe that the variability of renewable energy sources make it impossible for them to contribute significantly to electricity supply. This is mistaken, and reflects a lack of understanding of the way in which large integrated electricity supply systems are operated. It is regrettable that recent cost penalties applied to unpredictable supplies under the UK's New Electricity Trading Arrangements (NETA) have reinforced this common public perception.

An important characteristic of large electricity supply systems, indeed the key to efficient, low cost and reliable supply of demand, is aggregation. Attempting to supply individual consumers from dedicated (autonomous) supplies, whether conventional or renewable, is problematic due to the highly variable and almost completely unpredictable nature of their electricity usage. These are the challenges faced by stand-alone systems, and will be explored in more detail in Section 7. On the other hand, if large numbers of consumers are brought together, as when supplied by a common electricity supply system, much of the variation exhibited by individual consumers will cancel out. This statistical effect is referred to technically as diversity and the result is a much smoother demand characteristic that reflects the average consumer demand profile. A simple example[1] is shown in Figure 1 where the demand of a single consumer is compared with the aggregate of 10 such consumers whose demands are random in time. Note that in this example the loads are on average constant and do not reflect the typical daily demand shape; the significant effect of aggregation is nevertheless immediately apparent. When account is taken of the scale of modern electricity supply systems, with millions of customers interconnected, the importance of aggregation should be clear.

Even with aggregation, system load will vary through the day, and some of this variation is unpredictable. Highly sophisticated mathematical forecasting tech-

---

* Transmission systems, designed for the bulk movement of electricity, are operated at high voltage (generally above 200 kV), whilst distribution systems operate at a range of lower voltages right down to 400/230 V. Only large generator units such as coal and nuclear, and the largest of the off-shore wind farms, are connected directly to the transmission system. This is due to the high costs of high-voltage switch-gear and transformers.

[1] D. Millborrow, Integrating wind—economic and technical issues, *Proc. Wind Power Technol.*, short course, CREST, 2002.

Figure 1 Smoothing effect of demand aggregation

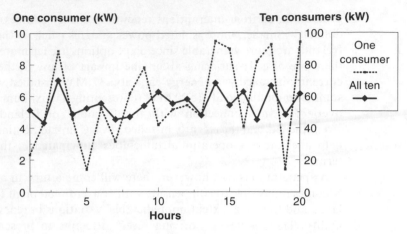

niques are applied to predicting the aggregate load a short time into the future so that the appropriate amount of plant capacity can be put on line, but inevitably so-called scheduling errors will occur. For the UK electricity system these are typically 1–2% (corresponding to around 500 MW). The system operators make sure that sufficient surplus capacity (often referred to as spinning reserve as it is usually held on part-loaded, *i.e.* spinning, plant) is available on the system.

The addition to the system of time-variable and somewhat unpredictable renewable energy capacity must be assessed in terms of any increases in scheduling error it may cause. However, substantial renewable energy capacity can be absorbed with no significant effect on the system, simply due to the existing fluctuations and uncertainties in demand. The Energy Review recently published by the Cabinet Office Performance and Innovation Unit (PIU)[2] concludes that up to 5% of the UK's electricity could be met from intermittent renewable sources with no additional system costs. For system penetrations of 5–10%, costs would add approximately 0.1 pence to the cost of a kW hour of electricity, and at 20% penetration the cost would be about 0.2 pence. These figures were based on analysis of wind power, which is very much a worst case. A renewable energy mix with substantial contributions from other technologies, such as tidal stream and wave energy, would almost certainly result in lower penalty costs.

It is also instructive to note that Denmark already meets 18% of its annual electricity supply from wind energy alone, and without any adverse impact on the safety and reliability of its system. However, the ability of the major Danish electricity suppliers to trade with neighbouring countries on the Nord Pool to compensate for fluctuations in wind power is an advantage not currently shared in the UK.

To date, no wide-scale system impacts from the integration of intermittent renewable energy sources have been identified to our knowledge. This is simply because the penetrations of renewable energy generators remain limited in all interconnected power systems. Large interconnected power systems, such as exist in Scandinavia, are highly robust and can tolerate significant uncontrolled

[2] *The Energy Review*, PIU report, 2002.

generation, as from intermittent renewable energy sources, without technical difficulty. Smaller, more isolated power systems (such as the one in Northern Ireland) are more vulnerable since plant options are far more limited. Concerns there are already surfacing about the impact of any further increases in the currently installed wind energy capacity (37 MW installed wind capacity *vs.* a summer minimum load of 520 MW representing a maximum penetration of just over 7%). Interconnection with the Republic of Ireland, which is to be strengthened, and the recently installed HVDC link to Scotland, are expected to help alleviate any operational difficulties anticipated in the near-to-medium term.

As penetrations rise, however, there will come a time in any system when it becomes impossible to compensate with conventional plant for the increasingly large, and to a great extent unpredictable, variations in renewable power availability. These issues are driving recent attempts to present wind energy as dispatchable, either through sophisticated control of power output or through combination with locally connected controlled generators. Only large-scale energy storage would allow unlimited penetration, but, aside from opportunities for large-scale pumped hydro installations (such as at Dinorwic, Wales), the technology is not sufficiently developed for this, and the economic implications are far from clear at this stage.

Some observers suggest that the development of the new technologies such as Regenesys and Superconducting Magnetic Energy Storage will open up low-cost, large-scale energy storage.[3] Others believe that hydrogen storage, as part of an extensive hydrogen economy, will banish load-matching issues and herald a completely sustainable energy supply system (provided the hydrogen is created from renewable energy sources). Hydrogen has long been held up as an energy panacea, but it would be rash to assume that this will happen in the near-to-medium term.[4]

Network integration issues

Steady-state Voltage Rise.   In general, the scale of the renewable energy plant (even when aggregated as in large wind farms) will be smaller than the large conventional coal, gas and nuclear power stations. As a consequence, this plant will normally be connected to lower voltage sections of the transmission/distribution infrastructure. When such electricity generators are connected to the system, they may cause local disturbances due to the limitations of the local network. In simple terms, the resistance of the network can result in voltage rise at the point of connection. Figure 2 shows the way in which the voltage on a supply line falls as additional local loads are added with distance from the local supply point. When a local embedded generator is added (as in Figure 3), the voltage is pushed up. For a well-designed system this will be well within the acceptable voltage limits. However, increases in generation capacity will in some senses have to be matched by increases in the current carrying capacity of the network and this has cost implications. Indeed, where these costs are borne is an

---

[3] *IEE Power Eng. J.*, 1999, **13**, 107–180.
[4] ETSU report F/03/00239/REP.

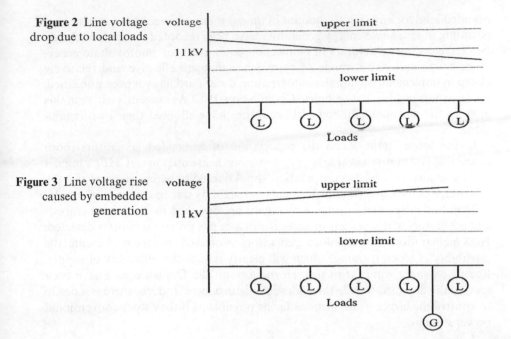

Figure 2 Line voltage drop due to local loads

Figure 3 Line voltage rise caused by embedded generation

important issue that is reflected in the contrast between 'shallow' and 'deep' charging for access to distribution systems.

Flicker. In addition to the steady-state voltage effects considered above, transient voltage variations reflecting short-term variations in the output of the embedded generator may occur. These changes can contribute to a measure of rapid voltage fluctuation known as flicker, so called because fast voltage changes will cause conventional incandescent light bulbs to flicker in intensity.[5] Transients caused by the connection and disconnection of generators can also contribute to flicker. Most electricity supply systems publish limits for allowable flicker and care must be taken that the installation of renewable energy sources does not result in these limits being exceeded.

Islanding. Operational safety must also be considered, although there has been a tendency for concern on this issue to be exaggerated. In particular, the issue of 'islanding' has been highlighted. Network operators in some EU countries require supply lines to be isolated before any remedial engineering work is allowed to take place on the network. Their concern is that these new and uncontrolled sources might energize a line that has been isolated from the main body of the system, thereby posing a danger to engineers working on the network. Sophisticated protection systems are thus stipulated, even though the probability of the renewable energy generator exactly matching the locally

5 CENELEC, *Flickermeter—Functional and Design Specifications*, EN 60868, 1993.

islanded load for any serious amount of time is minimal; indeed, no instances of islanding of renewable energy generators have been recorded to our knowledge. Nevertheless, manufacturers of connection equipment for photovoltaic generators have developed a range of approaches that are effective and relatively cheap to implement. Although no international standard has yet been published, the International Electro-technical Commission (IEC) are currently active in this area. In the meantime, different EU countries have adopted their own regulations.

In the longer term, when the penetration of embedded generators (both renewable and non-renewable such as combined heat and power, CHP) is high, it can be argued that islanded operation should actually be encouraged to take advantage of the possible improvements in reliability due to diversity of supply.

Traditional electricity protection systems are designed to disconnect embedded generators whenever a transient disturbance (*i.e.* potential fault) is detected. High penetrations of embedded generators protected in this way create the possibility of block tripping, which will clearly reduce the reliability of supply. Recent off-shore wind farms (*e.g.* Horns Rev off the Danish coast) have been specifically designed to ride through such disturbances. Indeed, there is a desire to control the larger wind farms as far as possible as if they were conventional power stations.

Harmonics.   Finally, and probably of least significance, is the issue of harmonics. All electricity supply systems exhibit some degree of waveform distortion. The departure from the ideal sinusoidal waveform can be measured in terms of the magnitudes of the harmonics of the fundamental. Any electrical generator will produce small harmonic currents and these will contribute to distortion of the voltage waveform. Very small generators, such as photovoltaic systems, often employ power electronic interfaces and these can generate potentially significant harmonics. Standards already exist for large generators, and these are currently evolving to cover grid-connected power electronic sources.

Potential Benefits of Distributed Generation.   Not all the consequences of embedded generation are negative. Because renewable energy generators are small and numerous in contrast to conventional generating stations, they can add to the security of supply, as highlighted by the PIU report.[2] Since they are distributed throughout the system they tend to be nearer to the loads than conventional power stations and consequently can result in a reduction in electricity distribution losses. Savings cannot be taken for granted and should be calculated. A simple calculation approach, known as the substitution method, is commonly applied in the UK. Load flow calculations are performed and the losses are calculated with and without the embedded generator connected. The resulting loss adjustment factors are used to aggregate up demand and generation to the location where the distribution network is connected to the transmission system. However, recent research has brought into question the accuracy of this approach.[6] Only a full probabilistic approach is capable of determining the actual

[6]   J. Mutale, G. Strbac and N. Jenkins, Allocation of losses in distribution networks with embedded generation, *IEE Proc. Generat., Transmis. Distrib.*, 2000, **147**, 1–14.

benefits, and as a result they are ignored in many countries. This is an area that would benefit from further study.

Additional potential benefit occurs when the installation of embedded generating plant avoids costs associated with centralized generation (such as expanding network capacity to meet increased local energy demand). In the USA, Pacific Gas and Electric (PG&E) analysed a small section of their network to investigate the impact of connecting in a 500 kW PV array.[7] Their results indicated benefits arising from avoided fuel costs, avoided costs of using a peak turbine, reduced ohmic and VAR (reactive power) losses, and avoided investment in expanding distribution capacity and associated maintenance costs.

It is clear from the above summary that embedded generation will have an impact on the network. Encouragingly, a recent study commissioned by the DTI concluded: 'there are no fundamental technical difficulties with the connection of increasing amounts of embedded generation, although in many cases there will be a cost'.[8]

3   Wind Power

Resource Implications

The wind speed at a given location is continuously varying. There are changes in the annual mean wind speed from year to year; changes with season (seasonal), with passing weather systems (synoptic), on a daily basis (diurnal) and from second to second (turbulence). All these changes, on their different time scales, can cause problems in either predicting the overall energy capture from a site (annual and seasonal), for wind speed measurements (synoptic, diurnal and turbulence) and for wind turbine design (all scales).

A typical plot of wind speed over a number of minutes is shown in Figure 4. It can be seen that there is considerable variation in wind speed over the whole period and that significant variations occur over periods of a few seconds.

Basic physics shows that the power available from a wind flow is proportional to the cube of the wind speed and so short-term wind speed fluctuations can translate into quite dramatic short-term variations in the output from a single wind turbine. However, wind turbines are commonly connected in substantial groups (wind farms) and fortunately the statistical effects of aggregation have a marked smoothing effect. If it is assumed for simplicity that the individual short-term fluctuations in output are uncorrelated between turbines, then the variability of the aggregate output will fall as $1/\sqrt{N}$, where N is the number of wind turbines in the group.

The UK's assessable on-shore wind energy resource has been estimated at over 300 TWh/annum, taking into account environmental restrictions and excluding sites with a wind speed of less than 7 m/s at 45 m above sea level.[9] As the technology gets cheaper, lower wind speed sites will become attractive; on the

---

[7] D. S. Shuger, PV in the utility distribution system: the evaluation of system and distributed benefits, *21st IEEE PV Specialists Conf.*, 1990, pp. 836–843.

[8] *Embedded Generation and Network Management Issues*, ETSU report K/EL/00227/REP, 2000.

[9] ETSU report R-122.

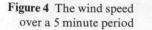

Figure 4 The wind speed over a 5 minute period

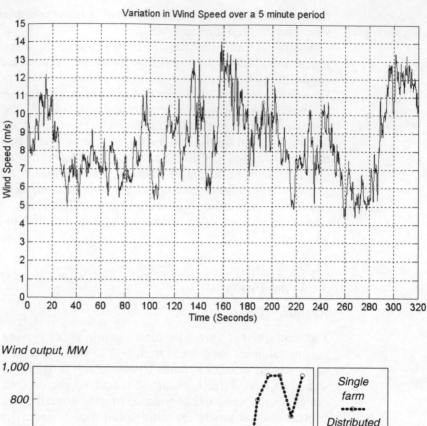

Figure 5 Impact of geographical diversity on wind power output

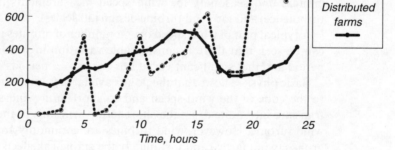

other hand, not all of the assessable resource will be commercially viable owing to high connection charges to cover local grid reinforcement and to availability of the land.

Geographical Diversity.   Weather variations are both temporal and spatial. For a maritime climate like the UK's, a number of different weather systems may be present at any one time, resulting in considerable variations in weather around the country. Wind speeds in particular will be affected. It has been known for some time that this will assist the integration of wind power at a national level.

Figure 5 contrasts a hypothetical 1000 MW wind farm with the same turbines distributed in groups around the country. It is evident that a high degree of

56

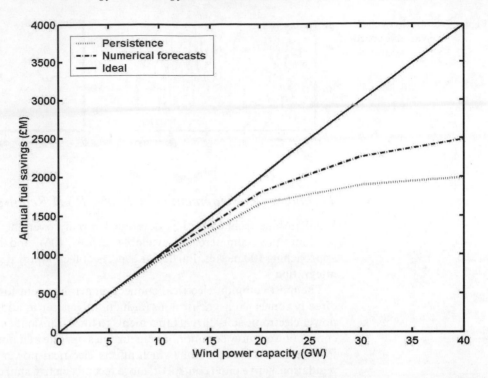

Figure 6 Value of wind power forecasting

spatial smoothing results from the geographically distributed arrangement.

In a study of five sites in northern Germany it was found that the coherence* between the sites falls quickly for frequencies greater than once per day.[10] This implies significant smoothing at time-scales relevant to integration into large power systems.

Wind Power Prediction. Clearly, if wind power can be accurately predicted, this will assist in the operation of the power system, thereby reducing operational penalties and increasing the value of the wind energy. A combination of statistical and meteorological models has been applied to this forecasting problem, and with a degree of success (Figure 6).[11] Building on this, the next generation of prediction tools is now evolving, which also make use of on-line data from wind farms.[12]

---

* Coherence is a statistical measure of the relationship between two time series as a function of frequency. A coherence of unity indicates identical variations at the two sites considered, and zero no statistical connection, at the frequency in question.

[10] H. G. Beyer, J. Luther and R. Steinberger-Willms, Power fluctuations from geographically diverse, grid coupled wind energy conversion systems, *Proc. EWEC'89*, Peregrinus, 1989.

[11] A. Joensen, G. Giebel, L. Landberg, H. Madsen and H. Nielsen, Model output statistics applied to wind power prediction, *Proc. EWEC'99*, 1999.

[12] G. Giebel, L. Landberg, T. S. Nielsen and H. Madsen, The Zephyr project, the next generation prediction system, *Proc. European Wind Energy Conf.*, WIP, 2001.

Figure 7 Smoothing effect of variable speed operation

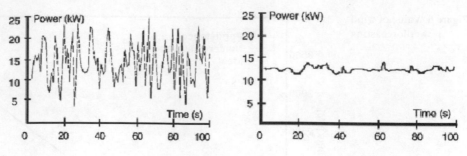

Technology Developments (Including R&D Requirements)

Wind turbine technology has developed rapidly over the last 20 years, with machines now commercially available up to 2 MW and having hub heights approaching 100 metres. Particular aspects of technology development relate to integration.

The most common electrical connection arrangement for wind turbines is a directly connected asynchronous (induction) generator; additionally, the use of power electronic soft-start arrangements to limit transient currents at the time of connection are now common. These turbines operate at a nominally fixed speed. Power control at the rotor (which affects electrical power quality) is by stall regulation, active pitch control or, more recently, active stall control. Driven by a combination of improved power quality and quieter operation, variable speed turbines are becoming increasingly common. They can also deliver higher energy yields by allowing the wind turbine rotor to operate under more favourable aerodynamic conditions for a greater proportion of the time. There is a cost to pay in the form of the power electronics (and their associated electrical losses) used to connect the variable speed electrical generator to the network, and they can contribute unwanted harmonics.[13] The main attraction of variable speed operation from an integration perspective is that the rotor inertia is available to smooth short-term variations in wind speed, thereby providing more constant power output and less contribution to voltage flicker. Figure 7 shows the effect of variable speed operation on the power quality.[14] The advantage of such smoothing is more pronounced for single turbine installations where the benefits of aggregation are absent. In principle, the inclusion of power electronics also allows the possibility of reactive power control that can be used to limit steady-state voltage excursions.

Most wind turbines employ a gearbox to connect the main shaft (which may be turning a slowly as 25–30 rpm on the larger machines) to a conventional 1500 rpm electrical generator. There is, however, increasing interest in direct drive arrangements in which a large multi-pole generator is coupled directly to the rotor. The main advantages of this arrangement include reduced gear noise (due to the elimination of the gearbox), simplicity and improved reliability. However,

[13] S. Heier, *Grid Integration of Wind Energy Conversion Systems*, Wiley, Chichester, 1998.
[14] M. R. Dubois, H. Polinder, J. A. Ferreira, Generator topologies for direct-drive wind turbines, *Proc. WECC 2001*, Matane, Canada.

power electronic interfaces are almost always needed for connection to the network. Therefore, in almost all cases these turbines will be operated at variable speed. Direct drive turbines are now available, or under development, from:

- Enercon (500 and 1500 kW)
- Lagerwey (750 kW)
- ABD Windformer (500 kW prototype)
- Genesys (500 kW prototype)
- Jeumont (750 kW)

Synchronous generators are used in these configurations; they can be externally excited, as with the Enercon designs, or permanent magnet.

Off-shore Issues and Technology

Offshore wind farms are relatively recent, with the first installation at Vindeby, Denmark, in 1992. Since then, further installations have been completed in the Netherlands, Denmark, Sweden and the UK and the total offshore installed capacity now exceeds 80 MW. Both Germany and the UK currently have ambitious plans for a major expansion of offshore wind energy, driven to some extent by the difficulties of obtaining planning permission for onshore developments.

Offshore sites tend to benefit from higher wind speeds and lower turbulence levels. In addition, because noise emissions are in most cases not of significant concern, the gearbox may be modified to allow a higher rotational speed than onshore, giving additional electricity production of up to 5%. However, installation is clearly more expensive than onshore; significant underwater cabling is required, and operation and maintenance is considerably more demanding. On balance, the electricity from offshore wind farms is currently somewhat more costly than onshore counterparts, by 5–50% depending on the location and installed capacity.[15] The expectation is that a combination of improved foundation designs (such as monopiles), cheaper installation techniques and larger turbines and wind farms will deliver electricity competitive with onshore developments. The UK's offshore resource is large, as would be expected for an island. If all possible sites were utilized, over 1000 GW of installed capacity would result (over 20 times the peak UK electricity demand). More realistic and realizable estimates suggest an annual yield of up to 100 TWh/annum.[16]

Connection Issues. The first generation of offshore wind farms have employed conventional medium-voltage three-phase AC connection. The balance between electrical losses within the wind farm and the costs of cables and switchgear is optimal at an operating voltage of around 35 kV. Middelgrunden uses 30 kV and the planned 160 MW wind farm at Horns Rev is planned at 33 kV. Since the operating voltage of the individual generators is less than 690 V, step-up transformers are required for each turbine; these are usually accommodated within

[15] J. R. C. Armstrong, Wind turbine technology offshore, *Proc. 20th BWEA Annual Conf.*, Cardiff, 1998.
[16] *The Energy Review*, PIU report, 2002.

the tower base. It would appear attractive to use higher voltage generators, but to date only ABB's Windformer technology does this (up to 20 kV DC).[17]

The first offshore installations have been sited reasonably close to shore, often less than 6 km. At these distances, transmission to shore *via* sub-sea cables at around 35 kV is acceptable and straightforward. As distances to shore extend, and as the installations become larger, it becomes more attractive to consider the installation of an off-shore sub-station to raise the voltage level to up to 150 kV. This is the approach planned for Horns Rev.

In future, high-voltage DC (HVDC) systems are likely to be used. For long cable runs it is expected to be cheaper than the AC equivalent, and it has the advantage (due to the power electronic interface) of allowing connection to weaker sections of the transmission network.

A range of technical possibilities exist once HVDC and its associated power electronics are considered. These include wind farms operating with DC generators (as with Windformer) and conventional fixed-speed turbines generating into a variable frequency AC wind farm network, connected to shore *via* HVDC. Further research is required to explore these different options.

4 Photovoltaics

Resource Implications

It has been estimated that if the total technical potential for building-integrated PV in the UK was exploited, this could generate over 266 TWh/annum (or more than the UK's current electricity needs of around 200 TWh/annum); if this is restricted to new build, the figure falls to only 7 million MWh/annum.[18] However, this same report accepted that only a small proportion of this would actually be installed, and that this would depend critically on the context. In contrast to wind, the diversity of the solar resource around the UK, and how this would affect the national electricity supply system, has not received significant attention.

Building-integrated PV is attractive from a number of standpoints. The PV modules can substitute for roof or façade elements, thereby reducing the net costs; they can be operationally integrated into the building; and they are unlikely to be the subject of public opposition in the way wind farms have been. A number of demonstration projects have been undertaken and these show how visually attractive such buildings can be. Figure 8 shows the prize-winning Doxford development in northeast England, where the entire south-facing façade is constructed using polycrystalline silicon PV modules, generating up to 73 kW.

---

[17] F. Ownman, Windformer—an integrated system solution for medium and large scale wind farms, *Proc. 2nd Int. Workshop on Transmission Networks for Offshore Wind Farms*, Royal Institute of Technology, Stockholm, 2001.

[18] *The Value of Electricity Generated from Photovoltaic Power Systems in Buildings*, ETSU report S/P2/00279/REP, 1998.

Figure 8 Interior and exterior views of the solar office at Doxford

Table 1 Record cell and module efficiencies under Standard Test Conditions[a]

	Cell efficiency (%)	Module efficiency (%)
Si (crystalline)	24.7	22.7
Si (multicrystalline)	19.8	15.3
CIGS	18.4	12.1
CdTe	16.4	10.7
α-Si	12.7	10.4
Nanocrystalline dye	6.5	

[a]Standard Test Conditions are 1000 V/m^2, AM 1.5 and 25 °C.

Technology Developments

PV products for terrestrial application have improved enormously over the last 15 years. This reflects the considerable research and development activities that have been undertaken over this period, most notably in the USA, Australia, Germany, Japan and the UK. Table 1 above summarizes the peak measured conversion efficiencies for the different available cell technologies, tested under laboratory conditions.[19] Commercial large-scale production is not expected to attain these peak values; the best commercial products currently have conversion efficiencies up to 17%.

Regrettably, the impressive improvements of module performance and reliability have not been matched by progress with the balance of system components. Inverters, particularly those for grid-connected PV, have been identified as a common cause of poor system performance.[20]

Surprisingly, the grid connection of PV has already raised issues of impact on the electricity supply system, despite the minimal penetrations. Electricity companies have raised concern about the potentially negative impact of grid connected inverters on power quality and safety, pointing out that the combined effect of numerous PV systems connected to the same section of the network might be significant. As a result, considerable research has been undertaken to examine

[19] M. A. Green, K. Emery, D. L. King, S. Igari, W. Warta, Solar cell efficiency tables (version 18), *Prog. Photovoltaics*, 2001, **9**, 289–293.

[20] G. H. Atmaram, Photovoltaic test laboratory accreditation and product certification, *30th IEEE PV Specialists Conf.*, 2002.

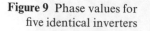

Figure 9 Phase values for five identical inverters

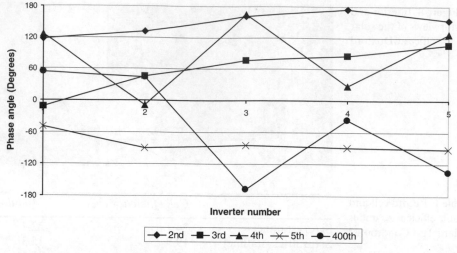

the way in which multiple inverters could interact and the expected aggregate impact on the local network.

Harmonic Aggregation. Power electronic converters, such as those used for the grid connection of PV, generate harmonic currents, which could on aggregate result in harmonic distortion of the voltage waveform of the local network. Figure 9 shows the phase relationship between the harmonics of five identical inverters connected to a common grid point.

The 400th harmonic in this example is caused by the high-frequency switching used in such power electronic devices, in this case to convert the DC from the PV array to 230 V/50 Hz AC for injection into the network. It can be seen that phases of the switching harmonic are more-or-less random, so a high degree of cancellation of this particular harmonic can be expected. This is clear from Figure 10, which shows the calculated impact of increasing numbers of inverters on the diversity factor,* a measure of the aggregate impact.[21]

The 3rd and 5th harmonics created by inverters tend to dominate, and it can be seen that the phases are fairly consistent across the different inverters, indicating that only limited cancellation can be expected. For these reasons, strict limits are being imposed on the harmonics allowed from inverters, primarily through the application of standards for grid-connected equipment.

Anti-islanding Protection. Utility concerns about the possibility of islanding (see Section 2) have led to the development of sophisticated anti-islanding techniques for inverters. A variety of approaches have been developed. Some rely on detection of changes to the voltage waveform associated with disconnection from the main supply system. However, the most popular are now based on a

---

* Diversity factor is a measure of the aggregate impact; 1 represents no effect, and 0 a perfect cancellation of the individual variations.

[21] D. G. Infield, Combined switching harmonics from multiple grid-connected single-phase inverters, *IEE Proc. Generat., Transmis. Distrib.*, 2001, **148**, 427.

Figure 10 Cancellation of switching harmonics with increased numbers of inverters

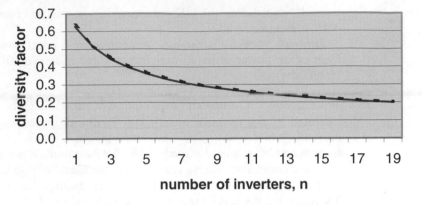

subtle manipulation of the generated current waveform, known as frequency shift. This is particularly simple and cost effective to implement through control of the inverter switching pattern to ensure that, on disconnection from the network, the generation frequency will rapidly drift out of the allowed operational range, forcing a trip. This approach has been demonstrated to be effective over a very wide range of network conditions.[22]

Economics. In contrast to remote area (off-grid) applications, where PV is often the least cost option, for grid-connected applications, electricity from PV is still relatively expensive, typically by a factor of 10,* depending on the level of the solar resource and cost of grid electricity supply. However, costs to the consumer are converging, aided by buy-down policies and other governmental support mechanisms, such as those offered in the UK, Germany, Japan, US and Australia. Most attention is currently focused on the reduction of costs of the PV modules. The building of larger automated continuous production units will drive down production costs, but in addition lower cost cell technologies are being explored. New low-cost devices such as those based on thin-film microcrystalline silicon and dye-sensitized technologies are under development and commercially attractive products based on CIGS and CdTe (thin film direct bandgap materials) are being developed.

In this drive to reduce the cost of the modules, it is important that the balance of system components is not overlooked.

5 Water Power

Conventional large-scale hydropower is a relatively mature technology, the potential for which in the UK has been fully exploited. Small-scale hydro offers some additional opportunities but is generally found to be expensive in the UK owing to the cost of the civil works, and also the turbine costs when heads are

---

* A factor of 10 is typical for the UK, but figures as low as 4 have been quoted for the US context.

[22] G. A. Smith, P. A. Onions and D. G. Infield, Predicting islanding operation of grid connected PV inverters, *IEE Proc. Elec. Power Appl.*, 2000, **147**, 1.

limited. Nevertheless, the UK's practicable potential has been estimated at over 3 TWh/annum.

The main hopes for substantial generation from water power in the UK come from ocean power in its various forms: wave power, tidal barrages and tidal stream devices.

Wave Energy. Commercial exploitation of wave energy has to an extent been inhibited by the plethora of different devices proposed, several of which are currently under trial.[23] The UK is well placed to exploit wave energy, with a huge accessible resource exceeding 600 TWh/annum. When account is taken of practical constraints, such as conversion efficiency, shipping lanes and environmental restrictions, the resulting practicable resource is estimated at around 50 TWh/annum, the bulk of this being far offshore.

The most popular shoreline device is the oscillating water column (OWC). With these designs, air is trapped in a chamber that is open to the sea below the water line. Changes in wave pressure on the opening cause the water level within the enclosure to oscillate, which in turn causes an oscillating air flow through a turbine. Conventionally this is a Wells turbine, which turns in the same direction irrespective of the air flow direction. OWCs are being demonstrated in the Azores,[24] the UK[25] and Australia. OWCs are not resticted to shoreline application and a floating system has recently been completed in Japan. A particular challenge for the electrical integration of OWCs is the pulsating nature of the power output. Attempts have been made to use flywheels to smooth the output, either on the turbine rotor shaft itself, or through electrical connections *via* power electronics. For multiple rotor systems, smoothing through aggregation is possible.

Tidal Barrages. Tidal barrages are a currently available technology, but very few exist worldwide. The best-known example is the 240 MW scheme at La Rance in France, and smaller installations have been made in Nova Scotia, Russia and China. The UK has a number of attractive sites due to its high tidal ranges, the largest potentials being on the Severn estuary (8600 MW capacity), and the Mersey (700 MW capacity). However, the scale of these installations and the calculated long payback periods make the required investments unlikely in the context of the privatized electricity supply industry (ESI). Furthermore, environmental considerations may present a barrier to large-scale developments.

Despite the relatively high costs, it may be that the time is right for government to consider whether such schemes would be an effective replacement for the UK's stock of ageing nuclear reactors. Although variable, the output from tidal barrages is entirely predictable, which is advantageous for integration into the electricity system.

Tidal Stream. Tidal stream technology, which directly exploits marine currents,

[23] T. W. Thorpe, *A Review of Wave Energy*, ETSU report R-72, 1992.

[24] A. F. O. Falcao, Design and construction of the OWC wave power plant at the Azores, *Proc. Wave Power: Moving Towards Commercial Viability*, ImechE, 1999.

[25] Heath and Whittaker, *The History and Status of the Limpet Project*, EU project report for SEAFLOW, 2002.

Figure 11 Indicative probability distribution for tidal stream output

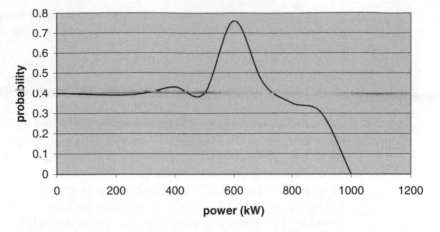

is relatively new but is presently generating considerable interest, especially in the UK. Turbine rotors can be used to extract energy from the flows, much as a wind turbine does, or alternatively oscillating aerofoils or hydroplanes can be used, as in the 150 kW Stingray device recently launched on Tyneside.[26]

The accessible UK resource has been estimated at around 36 TWh/annum for cost-effective sites with sufficient flow. However, it will take time and considerable engineering development to scale up and prove the technology before this resource can be exploited.

In common with barrages, the output will be variable, but to a great extent predictable.* Power output probability density distributions can be calculated based on tidal information and the energy conversion characteristics. These distributions can be surprisingly uniform, as in the indicative example shown in Figure 11, making tidal stream generation a reasonable approximation to base load plant. The addition of only a limited amount of energy storage is required to provide a true base load characteristic with constant power output.[27]

6 Biomass

Biomass is the only source of renewable fixed carbon, and closely resembles traditional fossil fuels in that it can be stored and used when required. However, unlike most fossil fuels, biomass is often limited by the energy density of the stored fuel. Therefore, it must be produced and consumed locally, as energy consumption associated with transportation over long distances might even exceed that of the fuel itself. This means that biomass power-generating units are relatively small compared to conventional plant (relying on local supply chains for feedstock) and possess the characteristics of small embedded generating units.

There are three basic thermo-chemical conversion technologies that use solid biomass as a primary fuel for the production of electricity, namely direct com-

---

* Tidal flows are also influenced to some extent by local weather conditions.

[26] *Prof. Eng.*, 2002, 10 July, p. 33.

[27] I. G. Bryden and D. M. Macfarlane, The utilisation of short term energy storage with tidal current generation systems, *Energy*, 2000, **25**, 893–907.

bustion, gasification and pyrolysis. In addition, the use of liquid biomass (such as sewage sludge) for the production of methane *via* anaerobic digestion is increasingly common.

Electricity production using solid biomass fuels is still a developing industry and as a consequence is not competitive on price with electricity from fossil fuels without some kind of government fiscal or policy support. However, it is competitive with nuclear power and possibly new-build clean coal power stations but not with modern gas-fired power stations within the current regulatory and economic climate. In the longer term, grid-connected biomass generation (using the full range of possible technologies) may become competitive; the greatest potential is for small-scale embedded generation using gasification, pyrolysis or high-speed steam engine-based plant. In the short term, small-scale (100 to 500 kWe) dedicated plants for use on farms or by rural industry has the greatest potential. In the medium term, when increased demand for electricity could be causing the grid to become overloaded and unreliable, then larger (1 to 20 MWe) embedded biomass generation plant providing end-of-grid support may become an attractive alternative to reinforcing the grid.

7 Small Grids and Isolated Systems

Smaller power systems, very often island-based, have provided the context for a number of the first demonstrations of renewable energy, primarily motivated by the higher local costs of conventional electricity generation. Island power systems range in size from hundreds of kilowatts to megawatts, and are usually supplied from diesel generator sets. Wind turbines have often been the most appropriate replacement technology, partly due to the generally good wind resource available on islands, but also due to the relative ease with which wind turbines can be connected to the electricity system, provided that the penetration levels are kept reasonably low.

Many examples of such systems exist, stretching back as early as 1982 at Block Island on the east coast of the USA.[28] Since then, wind turbines have been installed on Orkney, Shetland, Andros, Crete, the Azores, Gran Canary, and Lanzarote amongst many others.

Not all consumers are connected to central electricity supply systems. This is particularly so for the case of rural populations in developing countries. It has been estimated that more than 2 billion people presently have no access to electricity.[29] Despite vigorous efforts to extend the grid electricity supply system with an additional 1.3 billion successfully connected over the last 25 years, more people now lack electricity supply owing to population growth over the same period.[30] Renewable energy offers the possibility of least-cost small-scale decen-

[28] P. H. Stiller, G. W. Scott and R. K. Shaltens, Measured effect of wind generation on the fuel consumption of an isolated diesel power system, *IEEE Trans.*, PAS-102, 1983, 1788–1792.

[29] B. McNelis, G. van Roekel and K. Preisner, Renewable energy technologies for developing countries, in *The Future for Renewable Energy—Prospects and Directions*, ed. EUREC, James & James, 2002.

[30] J. Bond, Opening statement, *World Bank Energy Week: Extending the Frontiers of the World Bank's Energy Business*, World Bank, Washington, DC, 1998.

Figure 12 Wind diesel system configuration

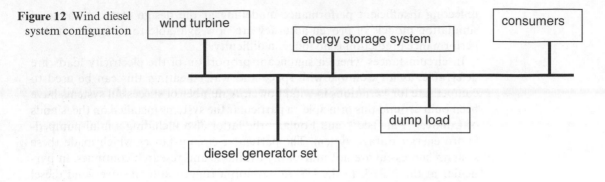

tralized electricity generation that may be suited to local conditions. For those who use as little as 2% of the energy consumed *per capita* in the West (and most of this in the form of firewood for cooking), relatively small quantities of electricity for domestic applications can make a significant difference in terms of health improvements, poverty alleviation and environmental impact.

Photovoltaics, in the form of solar home systems, are making the largest impact amongst non-grid-connected renewable energy forms, but micro-hydro and small-scale wind are also widely used, depending on the nature of local resources.

For stand-alone systems, the key challenge is to continuously match supply and demand. For the smallest systems, such as solar home systems, this is achieved using battery storage. This is acceptable when electricity costs are high and limited supplies are highly valued, as for domestic lighting, vaccine refrigeration or community applications, such as schools. For larger systems, the costs of batteries are prohibitive and combinations of renewables with small-scale conventional (often diesel) generators can be an attractive alternative. In scale, these systems sit between the solar home systems and the island power systems. A considerable market has been projected for these intermediate systems, in particular wind diesel systems in which a wind turbine is operated in conjunction with a diesel generator, with the system configured as shown in Figure 12. However, despite considerable research and development over the last 20 years, no commercially successful systems are available. This highlights the difficulty associated with the short-term matching of supply and demand and the lack of low-cost energy storage able to respond to the rapid variations of wind power associated with turbulence. Systems can be constructed without energy storage, but continuity of supply then requires the diesel engine to be run continuously to ensure a no-break supply to consumers.[31] This results in lower than expected diesel fuel savings owing to the poor part-load performance of small diesel engines.

A number of the existing wind diesel systems have been installed with grant aid, including most of those operating in developing countries. Their performance has often been disappointing, reflecting the lack of maturity of this technology. In many instances, initial expectations were unrealistically high, sometimes

[31] D. G. Infield, Wind diesel systems technology and modelling—a review, *Int. J. Renewable Energy Eng.*, 1999, **1**, 1.

reflecting insufficient performance modelling at the design stage. A range of simulation models of proven accuracy are now available to designers so that performance prediction is no longer a difficulty.[32]

In circumstances where a significant proportion of the electricity loads are deferrable, as for example with space and water heating, this can be used to compensate for variations in wind power. A number of successful systems have been built around this principle, in particular the systems installed on the islands of Lundy,[33] Fair Isle[34] and Foula,[35] the latter also including a mini-pumped-hydro energy storage system. The particular circumstances which made these systems successful are not universally available and research continues, in particular at the NREL in the US, to develop a rugged and effective wind diesel technology for general application.

[32] J. F. Manwell, A. Rogers, G. Hayman, C. T. Avelar and J. G. McGowan, *HYBRID2—A Hybrid System Simulation Model, Theory Summary*, NREL report XL-1-11126-1-1, 1997.

[33] D. G. Infield and J. Puddy, Wind powered electricity generation on Lundy Island, *ERIC IV*, ed. J. W. Twidell, Pergamon, Oxford, 1984.

[34] W. E. Stevenson and W. M. Somerville, The Fair Isle wind power system, *Proc. 5th BWEA Workshop*, Cambridge University Press, 1983.

[35] W. M. Somerville and D. Alexander, Experiences of a hybrid wind turbine hydro generator system with pumped water storage, *Proc. Rhodes Workshop*, June 1992.

Landfill Gas and Related Energy Sources; Anaerobic Digesters; Biomass Energy Systems

ADRIAN LOENING

1 Introduction

This review is concerned with the potential for power generation from freely available organic material. Although mostly concentrated on the production and use of methane by biological processes, it also examines the physicochemical production of other liquid and gaseous biofuels from organic feedstock. Much of the technology for this form of renewable energy could be classified as 'Energy from Waste'; however, it should be noted that this would also include the direct incineration of, or fuel production from, waste materials which are fossil in origin. Production of energy from sources such as the incineration, or gasification, of municipal solid waste is therefore not considered in this article, as much of the calorific value comes from the plastics content of this waste stream. Similarly, the use of biomass in direct-fired applications is not considered. It is, of course, recognized that much of the world relies on direct use of biomass for heating and cooking, mostly timber and dung; however, this use of biomass is outside the scope of this review.

The use of methane as an energy source is divided into two broad technical definitions: the collection for use of fugitive emissions of methane and the deliberate production of methane in closed processes. It could be argued, and the author would agree, that there is little environmental difference between the collection of fugitive emissions of methane from biological sources and those from geological, or fossil, sources, both having an equal impact on the environment. However, it has been generally accepted that methane captured from, for instance, offshore drilling operations would not qualify as a 'renewable resource'. A question in this definition must hang over methane from coal workings or natural seepage from coal seams, as either the result of human actions or natural processes. However, this article concentrates solely on energy production from

Issues in Environmental Science and Technology, No. 19
Sustainability and Environmental Impact of Renewable Energy Sources

Figure 1 Typical landfill gas production curve (Dom Eq = domestic equivalent waste stream)

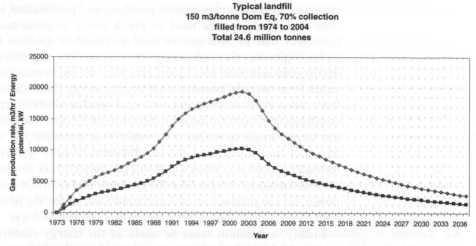

Typical landfill
150 m3/tonne Dom Eq, 70% collection
filled from 1974 to 2004
Total 24.6 million tonnes

inert wastes) will be made in view of other engineering and environmental factors that may affect the rate of gas production or the ratio of methane to other gases. A number of models have been developed over the last few years, but all are based on first-order decay mathematics, with variables that allow the adjustment of rise and decay rates for each waste type. Although the theoretical quantity of gas from a tonne of typical domestic refuse is around 500–600 m^3 at 50% methane concentration, the conditions within any given landfill site are unlikely to be ideal. The computer model is therefore adjusted by the empirical experience of the landfill gas project developer, and figures between 150 and 250 m^3/tonne are more typically used. Other empirical adjustments will be made to take account of moisture content, collection efficiency, depth, ambient temperature, aspect ratio (surface-to-volume ratio) and other site-specific information. The output of a gas model provides the starting point for development of a project, with a graph of gas and energy availability against time (see Figure 1).

Following the development of a model, it is always necessary to carry out trials on the site, consisting of the installation of a gas collection system and flare stack and operation of this equipment for a typical period of three months. Through this process it has been shown that although, on average, the various computer models are fairly accurate in their prediction of gas generation rates, there can be an immense variation in production from one site to another.

The reasons for these variations are complex but certainly include factors such as water content, level and availability, as well as temperature. Other factors such as movement of bacteria and 'seeding' of the waste mass with anaerobes clearly have effects and sites which have accepted wastes from other anaerobic processes, such as sewage sludge, will have greater breakdown rates than normal. Factors such as the presence of bacteriocidal compounds must also be included and a chemical analysis of the leachate (the liquor produced by the infiltration of rain water) will provide some indication of the biological condition of the waste mass.

Table 1 Estimated world-wide deployment of landfill power-generation projects

	Current capacity (MWe) By 1996	Estimated capacity (MWe) By 2010
World-wide	1385	4529
EU + EFTA	573	1577
CEE	0	25
CIS	0	225
NAFTA	730	2326
OECD Pacific	30	199
Mediterranean	0	4
Africa	2	23
Middle East	0	10
Asia	30	91
Latin America	20	49

World estimates of energy available from landfill gas vary up to a figure of about 9000 MWe (megawatt equivalent) of electrical energy potential. The European Commission suggests that at the current rate of development a capacity of around 4500 MWe will be installed and operating by 2010 (see Table 1).[3] However, the author believes that this underestimates the contributions for both Asia and Latin America.

Biomass

Fuel feedstocks for production of energy through anaerobic digestion, pyrolysis and gasification are available from two main sources. Waste organic material from the food processing industry and domestic waste are available where separate collection systems have been instituted; however, the main fuel resource is in the deliberate agricultural production of energy crops or the collection of waste materials from agricultural or timber production processes.

A large range of crops can be used for energy production purposes, the availability and potential of these crops being largely a function of soil types and weather. In general, and in particular in the EU under the set-aside scheme, energy crops would be grown on land that is either not needed or unsuitable for food production. Various studies have outlined the possible land areas that fall into these categories and current estimates range from 1.0 million ha to 5.5 million ha in the UK. The US Department of Energy estimates that around 77 million ha of land could be available to produce energy crops. The energy produced from these crops can be utilized in a very wide range of processes, from the production of biodiesel for road fuel use, through direct heating or combined heat and power application, to the production of electricity by fuel cells supplied with hydrogen from gasification processes. The efficiency with which each of these conversion technologies operates is very variable and the energy available from 1 ha of crop is both species and latitude dependent. Table 2 shows the gross

[3] European Union online web site, http://europa.eu.int/comm/energy_transport/atlas/assets/lgpostw.xls; search date October 2002.

A. Loening

Table 2 Gross energy potential of crops[4]

Fuel	Heating value (MJ/kg)	Approximate crop yield (T/ha)[a]
Lignin	28.0	
Wheat straw	14.5	6.0
Energy cereals	15.0	12.0
Miscanthus	17.0	30.0
Wood	17.0	10.0
Eucalyptus	19.53	10.0
Bamboo	15.85	
Rape straw	17.0	8.0
Heating oil	41.0	

[a]Approximate figures as individual locations vary widely.

Table 3 Gross energy potential of oil-producing crops[4]

Fuel	Heating value (MJ/kg)	Approximate oil yield (T/ha)
Sunflower oil	35.3	0.88 to 1.67
Rape oil	37.2	1.26
Linseed oil	36.9	
Palm oil		7.8
Olive oil		0.4 to 5.0
Diesel fuel	38.4	

energy potential of crops suitable for direct combustion, gasification or pyrolysis and Table 3 those oil-producing crops suitable for the production of diesel substitute fuels. It should be noted that the oil-producing crops also produce a waste biomass stream in the form of crushed seeds from which the oil has been extracted, most of which is currently used as animal feed, and the straw mass which is composted or ploughed back into the soil.

While agricultural production of crops suitable for energy use is immense (the FAO 1995 estimate of harvest, yield and production in Europe and the Mediterranean region exceeded 165 million ha under agricultural production of crops ranging from cereals to ground nuts), it has been shown that the economics of energy production require that feedstocks are drawn from fairly limited areas in the vicinity of the energy plant. Typical biomass-to-energy plants draw on agricultural and timber residues for a radius of 80 km around the plant. The US Department of Energy suggests that, within this radius, if energy crops were planted on approximately 4% of the land (approximately 20 100 ha) sufficient fuel for a 100 MW biomass power plant could be produced.[5]

[4] *Potential Energy Crops for Europe and the Mediterranean Region*, available from http://www.fao.org/regional/europe/escorena/b46/cont.PDF.
[5] *Biopower Feedstocks*, US Department of Energy information web site, http://www.eren.doc.gov/biopower/feedstocks/index.htm; search date October 2002.

Waste Bio-oils

Some mention should be made of the energy potential of waste or second-use oils. Two distinct sources of waste oils are produced by the food processing industry: used vegetable oil (UVO), primarily from the restaurant trade, is currently collected for use in animal food in the UK and much of Europe. Current UK estimates indicate that around 80 000 to 100 000 tonnes of this material is collected by over 200 small companies and refined and traded by a small number of specialist recyclers. Recent developments in legislation may remove this oil from the food chain, making much of the material available for power generation or road fuel use.

Similarly, considerable quantities of tallow are produced from the rendering of animal carcasses and, while much of this is used in direct heating applications for the rendering process itself, a certain amount is available for other uses. In the UK, as a result of the policy developed to control the spread of BSE, a substantial stockpile of this material has been accumulated, currently estimated at 220 000 tonnes. While not a huge resource, a continued stream of tallow and UVO will become available for energy use.

3 Fuel Production Technology

Landfill Gas

The technology for the collection and treatment of landfill gas has developed slowly over the last 20 years. In essence, all collection systems operate on the same principle, but the exact engineering detail depends largely on the engineering of the landfill itself.

It is almost always the case that a landfill will be engineered to suit the purposes and legislation of the waste management industry and not the production of energy. Indeed, the author would not recommend that any landfill site was developed specifically for the purpose of energy production. Gas is therefore collected from a wide range of landfill sites, from the traditional 'dump' or 'tip' where largely uncontrolled waste is filled into a void, to fully contained 'sanitary' landfills with engineered sealing systems.

Although a number of different approaches to the collection of landfill gas have been tried, it is generally accepted that the most successful collection systems are installed on completion of filling and sealing of individual 'cells' within the whole site. Vertical gas collection wells are drilled into the waste mass (see Figure 2) and these are connected to a series of horizontal collection pipe (usually made of MDPE) which are laid on, or buried just below, the surface of the site.

Gas is drawn from the site by a low-suction, high-flow rate, gas pump (typically a centrifugal blower) and the flow rate from each well is controlled by a series of valves. In the absence of any gas extraction system the generation of gas produces a slight positive pressure within the site and this will drive the gas out of the waste mass, resulting in fugitive emissions. This process has led to some well-documented disasters in the neighbourhood of landfills where gas migrated out of the site and accumulated and ignited in buildings.

Figure 2 Drilling for
landfill gas

In order to prevent this migration of gas, the objective of the gas collection system is to maintain a pressure slightly below atmospheric within the waste. Thus the collection of gas acts to reverse the migration direction. Even on the most highly engineered landfills the waste mass is not entirely sealed from the atmosphere and a negative pressure will draw air into the site where it will be mixed with gas and could ultimately destroy the anaerobic conditions. The operator must therefore carefully balance the need for adequate suction with the oxygen concentration in the resulting gas mixture.

A landfill can be considered an uncontrolled anaerobic digester, the operator having little control over the anaerobic processes. A number of techniques such as recycling of condensate from the gas or leachate from the waste mass can have an effect on the rate of degradation of the waste, particularly in very dry sites.

Biomass: Pyrolysis and Gasfication

Two basic processes are used for the conversion of biomass into fuels that are more easily handled, stored and combusted. Gasification involves the heating of organic materials in the presence of a controlled quantity of oxygen and water, the basic reactions being $C + H_2O \rightarrow CO + H_2$ and $C + \frac{1}{2}O_2 \rightarrow CO$. The first of these reactions is endothermic and the second exothermic. The process of gasification has been known for a considerable time and the production of 'town gas' from coal has been utilized since the Victorian era. The gaseous mixture resulting from gasification of fuels containing large quantities of carbon is a combination of carbon monoxide, hydrogen and methane, as well as a number of longer chain hydrocarbons collectively known as 'Syngas' and can be used in a wide range of applications.

Pyrolysis is a high-temperature process, where the biomass feedstock is heated in the absence of oxygen, generating vapours and some charcoal. The vapours are cooled and condensed, forming a mobile liquid, usually dark brown or black

	Physical property	Typical value
Table 4 Typical analysis of wood-derived crude pyrolysis oil[5]	Moisture content	15–30%
	pH	2.5
	s.g.	1.2
	C	56.4%
	H	6.2%
	O	37.3%
	N	0.1%
	Ash	0.1%
	Heating value	16–19 MJ/kg
	Viscosity	40–100 cP
	Solids (char)	1%

in colour. The exact properties of this liquid vary considerably, depending on the type of feedstock and the temperature and retention time within the pyrolysis reactor. Typically, the non-condensing gases produced by the pyrolysis process are used directly in the plant to provide the heat required to operate the process.

Pyrolysis oil typically contains around 15–30% water content and attempts to remove this by distillation causes other chemical reactions, forming chars which may be up to 50% of the weight of the original oil.[6] Pyrolysis oil is therefore considered to be chemically unstable and needs to be stored at relatively low temperatures. Even at room temperature, in certain pyrolysis oils there is a slow chemical change. Pyrolysis oils typically have a calorific value of *ca.* 17 MJ/kg, compared with fossil oils with a calorific value of 42–44 MJ/kg and, owing to their water content, are not miscible with other oils. Table 4 shows the typical analysis of wood-derived crude pyrolysis oil.[7]

There are a wide range of commercial gasifier technologies available, these having been developed since the 1920s for the production of town gas and the recovery of sulfur. Currently, there is considerable interest in the development of processes that will accept a wider range of feedstock, such as sorted municipal sold waste (MSW); however, this feedstock represents a number of distinct problems for the technology as the energy content is typically low and the water and other contaminant content high.

Recent developments in this technology for the processing of both biomass and MSW typically use a combination of pyrolysis and gasification technologies. In all of these processes the feedstock stream is compressed to a high degree and heated to drive off volatile gases, forming a carbon-rich char which is fed to a relatively conventional gasification process. Generally the gas production is used to provide the process heating required for the pyrolysis stage and it is the carbon monoxide and hydrogen that are left for energy use.

Within Europe and the US there are a number of small- to medium-scale

[6] A. V. Bridgwater, *Fast Pyrolysis of Biomass of Fuels and Chemicals*, Aston University, UK, March 1999.
[7] A. V. Bridgwater and G. V. C. Peacock, *Fast Pyrolysis Processes for Biomass*, Pergamon Press, Oxford, 1999.

projects for the conversion of both waste and harvested biomass. These systems almost always consist of a combination of pyrolysis and gasification. In most of these projects the gas is used to raise steam and not for direct combustion in spark ignition or diesel engines. Early trials with engine technology have shown that both types of pyrolysis oil are very difficult to combust in reciprocating engines, although there is a long-term objective to achieve this. There is some development of gasification processes with the aim of generating hydrogen fuel for use in transport applications. While direct combustion of hydrogen in internal combustion engines is inefficient, the eventual development of commercial fuel cells, directly generating electricity with high efficiency from hydrogen, holds some hope for the longer term solution to vehicle emissions.

The ARBRE project,[8] currently being developed in the UK, takes a feedstock of coppiced willow for gasification. The project is intended to generate around 8 MWe of electrical power using the gas produced in a combined cycle gas turbine. The first plant is projected to require feedstock taken from 2800 ha of planted coppice.

In most of these processes, with the exception of the ARBRE plant, the main driving force is the requirement to treat domestic waste using processes other than direct incineration or landfilling, and all of these processes rely on a gate fee, rather than the energy produced, to provide economic returns. There is, however, a great deal of legitimacy in this approach as both waste disposal and energy production are pressing environmental concerns.

Biomass: Anaerobic Digestion

Where the biomass feed stock has a particularly high water content, or is composed mainly of food or animal wastes and conventional sewage sludge (rather than woody wastes), processing by anaerobic digestion may be the preferred option.

A typical anaerobic digester accepts a feedstock of up to 10% dry solids (although some processes may be capable of handling up to 40%). This is inoculated with anaerobic bacteria and held in a tank for a period while digestion takes place. Typically, some mixing and recirculation of the liquors produced is required to ensure complete digestion. In addition, the process requires that the temperature is maintained at around 35 °C for the slower mesophilic bacteria and up to 55 °C for the faster thermophilic bacteria.

These processes are essentially the same as are occurring in a landfill, but the speed of digestion is increased as a result of the higher organic content of the feedstock and the improved mixing and inoculation of the process. In addition, as this is a closed system, 100% of the resulting methane can be collected and ingress of oxygen is prohibited.

Other than conventional digesters operating on domestic sewage plants, many of which use the methane collected for on-site combined heat and power, there are a number of projects in Europe that have been developed for the digestion of a wide range of feedstocks. These include the Valorga process for organic wastes

---

[8] Information web site of the UK ARBRE Project, www.arbre.co.uk; search date October 2002.

developed in France and currently being used in Amiens to process around 56 000 tonnes of organic refuse per year as well as at 11 other European sites. The Kompogas[9] system, using a thermophilic process to digest food wastes, is used in around 18 small-scale plants in Germany and Switzerland. In the UK the Holsworthy project[10] uses slurry from local farms as well as waste from food processors to produce methane which is used in a combined heat and power scheme. The project will initially provide 2 MWe of power to the grid and the digestate, left over following the completion of digestion, will be returned to the local farmers as a pathogen-free fertilizer. It is important to note that anaerobic digestion, particularly thermophilic processes, has the major advantage of providing high quality fertilizers which have very low pathogen counts. It is inevitable that this fertilizer is required to maintain soil fertility and condition and must be considered as part of any sustainable biomass energy cycle.

Bio-oils

Most bio-oils do not need any pre-processing before use other than filtration, which is achieved with standard centrifuging techniques. However, unprocessed oils typically are unsuitable for combustion in processes other than direct heating or steam raising.

For easy compatibility with modern vehicle diesel engines, bio-oils are converted into biodiesel. Biodiesel is produced by a process called 'transesterification'. The vegetable oil (or animal fat) is first filtered and then chemically reacted with an alcohol (normally methanol) in an alkali solution (normally sodium hydroxide) to produce esters. The process can also be achieved using ozone and water in a more complex reaction (but with easier to handle reagents). It is these esters that are collectively known as biodiesel. Glycerol is produced as a by-product and this can be used for the manufacture of soap.

The process of manufacturing biodiesel reduces the energy content by as much as 15%, with this 'lost' energy remaining in the glycerol. However, for road-fuel use this is acceptable as biodiesel is a direct replacement for, and mixable with, petroleum diesel.

It should be noted that biodiesel produced using the methanol process cannot be classed as a 100% renewable fuel as considerable quantities of methanol (normally from fossil sources) have been used in production and are still contained, albeit chemically altered, in the biodiesel.

Although biodiesel is routinely produced for addition to fossil diesel in France and is increasingly used for road transport fuel in Germany, the scale of production is low and in the UK the taxation of road fuel, despite recent concessions, makes this use of bio-oils relatively unattractive.

---

[9] Information web site of Kompogas AG, Switzerland, www.kompogas.ch; search date October 2002.

[10] Information web site of the Trade Association of UK Bioenergy Industry, http://www.britishbiogen.co.uk/bioenergy/21stcenturyfuel/bionrgmoppor.html; search date October 2002.

4   Conversion Technology

Early uses of landfill gas included the direct firing of brick kilns and proved to be very successful but was constrained, by the difficulty of compressing and transporting the fuel, to a few landfill sites that were in close proximity to brick or cement firing processes.

Since 1984, substantial use has been made of landfill gas in the UK for the production of electricity and as the preferred technology this has proved to be the most popular. On most projects up to around 20 MW, operators prefer the use of reciprocating spark ignition engines; on larger projects, though, the use of gas or steam turbines may become economically viable, particularly where there is a use for the pass-out steam or waste heat produced from this equipment. Currently, there are only a few combined heat and power projects based on landfill gas as, like the landfill itself, landfill gas schemes tend to be located in less densely populated areas.

Typical power generation projects now use a number of smaller, modular engines, although, on some of the larger projects, dual-fuel engines (using up to 9% of diesel fuel) have provided reliable operation. Considerable improvement has been made in spark ignition gas engines with the introduction of lean-burn technology and higher speed turbocharged engines and this has generally removed the requirement to clean and compress the gas to the previous levels. These developments have been driven by the EU landfill directive, which requires that gas shall be collected and where possible used for power generation,[2] resulting in an increase in the sale of gas engines to several hundred per year in the EU.

The engines currently being used for almost all landfill and biogas applications are supplied as fully containerized units ranging from 300 kWe to 1.4 MWe and are complete with all the required control systems and synchronization equipment needed for connection to the grid. Power is usually delivered from direct-drive generators, implying that the engines operate at 1500 rpm for 50 Hz supply and 1800 rpm for 60 Hz supply, although in a few cases lower-speed engines fitted with a gearbox can also be used.

Modern engines operating on gas from landfill sites or digesters provide one of the few truly base-load renewable power sources.

5   Constraints

The use of both landfill gas and pyrolysis oils or gases in reciprocating engines implies a considerably higher cost of maintenance than either natural gas or biogas from digesters.

Although many of the maintenance problems associated with landfill gas have been overcome, largely through experience and improvements in engines, the combustion of pyrolysis fuels still faces severe problems. Over the last decade, experience of operation of landfill-gas engines runs to many millions of hours, while the few engines operating on pyrolysis oil or gases can be counted in hundreds of hours.

Landfill gas is estimated to contain over 350 trace compounds, many of which,

Table 5 Typical composition of landfill gas[11]

Analyte	Maximum concentration	Minimum concentration
Methane	58.9% vol	43.8% vol
Carbon dioxide	39.7% vol	32.9% vol
Oxygen	3.09% vol	0.18% vol
Hydrogen	0.08% vol	<0.01% vol
Moisture	4.11% vol	1.26% vol
Total S	430.5 mg/m^3	30.8 mg/m^3
Total F	20.3 mg/m^3	5.6 mg/m^3
Total Cl	77.9 mg/m^3	14.7 mg/m^3
Organo S	<199.9 mg/m^3	<8.0 mg/m^3
H_2S	400 mg/m^3	21.4 mg/m^3
Organo Si	148 mg/m^3	21.4 mg/m^3
CO	146 mg/m^3	22 mg/m^3
Total NMVOC	1440 mg/m^3	<120 mg/m^3

even in small quantities, produce combustion by-products damaging to engine sets. Indeed, it is the presence of these contaminants that has all but mitigated against the use of gas turbines in landfill applications. Table 5 shows the contaminants found in a range of landfill gas samples taken by LQM Ltd in the UK.[11] There are a number of issues surrounding the combustion of gas with these contaminants, each of which requires different solutions.

The hydrogen content of landfill gas can be particularly high in gases generated from fresh waste and the ease of combustion of this element often causes pre-ignition in spark ignition engines, implying that the engine should be operated at a lower efficiency or output. In addition, volatile halogenated compounds, volatilized from waste-containing solvents, appear in higher concentrations in the early gas given off by a landfill. It is for this reason, and to prove the gas quantities, that most operators prefer to flare off the gas from newer areas of the site prior to utilization in engines.

Both the halogen and sulfur compounds form acids during combustion and these react with the bearing surfaces within the engine as well as with the lubrication oils. One of the major developments in the reliability of landfill-gas engines is the use of oils with high alkalinity or TBN (total base number). This alkalinity or ash content of the oil has increased the oil life from, in some cases, a few hours to a typical 1000 hours of operation on engines with extended oil capacity. However, great care must be taken not to have too large an ash content as this in itself produces deposits in the combustion chamber.

Since the removal of boron (in the form of borates) as a whitening agent in detergents, and its replacement with silicon compounds, the quantity of organic silicon in landfill gas has been increasing. This is particularly apparent on landfills which receive sewage sludge. These silicon compounds form oxides which tend to coat the inside of the combustion chamber. It is these deposits (see Figure 3), on some projects, that are the main reason for premature failure of engine components.

[11] R. R. Gregory, *Development of a Sampling Protocol and Standards for UK Landfill Gas Generating Set Emissions*, Land Quality Management Ltd, Nottingham, UK, 2002.

Figure 3 Engine
contamination

Bio-oils and Pyrolysis Oil

The use of bio-oils, in particular UVO and tallow but not biodiesel, in converted diesel engines results in an incomplete combustion and rapid coking of the combustion chamber. Cleaner virgin oils such as rape oil will burn more successfully but often result in a very high particulate load in the exhaust gases.

To produce a satisfactory combustion of these fuels there are current, patented, technologies which enhance the combustion temperature through the injection of small quantities of oxygen with the combustion air. At the time of writing, a considerable amount of research work has been done in this area but the prohibitive cost of the oxygen supply means that this is not yet a commercially viable technology. Advances in gas separation membranes holds out some hope that the direct use of bio-oils will become viable in the near future.

Because of their water content and inherent lack of lubricity, the combustion of pyrolysis oils in reciprocating engines has proved difficult to achieve. Initial trials have achieved a maximum of only 10 hours of operation of a converted diesel engine. These tests were not long enough to develop maintenance standards and showed that the exhaust emissions were outside accepted standards. Ormrod diesels and Aston University[12] have operated a 230 kWe six-cylinder engine on a dual-fuel basis. This engine was fitted with two fuel pumps per cylinder, allowing fossil diesel to be injected first, causing ignition of the pyrolysis oils. These trials showed exhaust emissions containing particularly high CO output, although the NO_x content was low. This indicated an incomplete combustion was taking place, although the reasons for this are not yet clear.

Oxygen enhancement technology has been used in a recent study[11] and this has shown that low exhaust emissions can be achieved, although owing to the

---

[12] J. Blowes, *Evaluation of Complementary Technologies to Reduce Bio Engine Emissions*, ETSU document number B/T1/00761/00/REP, May 2002.

Figure 4 Landfill gas equipment costs *vs.* plant capacity

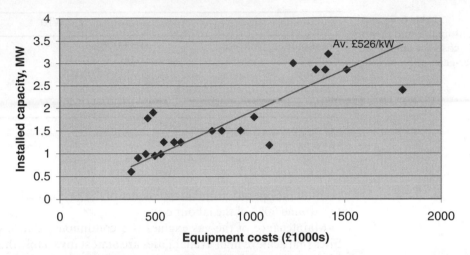

higher combustion temperatures the level of NO_x emissions would require the use of an abatement system.

Both UK trials of combustion of pyrolysis oil in engines ended prematurely with failure of the fuel-inject system, underlining the difficulty of using this fuel in applications other than direct heating.

6 Economics

Landfill Gas

Installation costs for landfill gas projects have fallen in line with both the significant competition in the gas extraction equipment market and the increased production rate for engines in the 1 to 1.3 MW range.

Typical equipment costs *versus* capacity for a small sample of landfill gas power generation projects in the UK that the author has been involved with are given in Figure 4 and indicate an average cost of £526/kW for equipment only. The projects from which this financial information was derived include ones installed under all the NFFO and SRO contracts and the predicted cost of projects yet to be commissioned under these schemes. The data also include some projects that are an extension of existing installations. The installed equipment costs do not include the costs of engineering or financing landfill gas projects but represent only the costs of engines, gas extraction systems and civil engineering. It should be noted that some of the cases studied were developments on sites with existing gas extraction systems that have been transferred to the project developer, reducing the capital costs.

Development costs of landfill gas projects outside the EU or the US, which are the largest markets for this technology, are likely to be greater as the market for equipment and qualified contractors is not so large.

Operational costs remain the major factor in the economic performance of landfill gas projects. Rigorous maintenance procedures must be in place to ensure the reliable operation of the projects. On all but the smallest closed

Table 6 Typical consumable and labour costs for landfill gas power-generation projects

Maintenance item	Average cost (kWh)
Oil including analysis	£0.00076[a]
Parts	£0.0035–0.0041[b]
Labour	£0.0055–0.0085
Insurance and administration	£0.0008–0.0013
Total	£0.0105–0.0146

[a]At oil price of £0.75/L.
[b]At €1.61/£1.

landfills the operation and maintenance of the gas collection system will account for around 30% of the labour costs.

Maintenance of the gas engines is a continuous, almost daily, requirement. Sites that achieve high availabilities are almost invariably those with permanent engineering staff. Table 6 shows the typical parts, oil and labour requirements for a modern landfill gas engine averaged over a 20 year life span. It should be noted that this table does not include data for investment, management and depreciation costs. Staffing requirements vary considerably from operator to operator, some preferring lower availabilities and remote operation while others choose to provide full-time staff, including 24-hour call-out facilities.

In many cases the owner and operator of the landfill gas plant will not be the owner of the fuel supply and it is normal practice in these cases for the power generation project to pay a royalty to the landfill owner in addition to assuming the investment costs and responsibility for gas collection and control.

Biomass Technologies

As most biomass gasification plants are currently in a development or prototype stage, estimates of both construction and operation costs vary greatly. Typical demonstration plants have construction costs of between £1800 and £3000/kWe and it is likely that these will reduce with the increase in numbers and scale of the plants.

Cost data for the Arbre project, at the design stage, taken from the ETSU New and Renewable Energy Report,[13] are shown in Table 7. These are extrapolated for full-scale operation on the assumption that a 33 MWe plant would be constructed. The recent liquidation, albeit possibly temporary, of the Arbre project underlines the difficulty of financing large-scale, renewable energy systems, particularly those that rely on multiple fuel sources and volatile fuel pricing.

The Arbre project differs from many of the other gasification systems mentioned in the previous section in that it was primarily seen as a power production technology and as such includes the cost of its fuel. Most other gasification technologies are aimed at using a waste product as the primary feedstock and attracting a gate fee for the disposal of these wastes. As such, the economics of

[13] Energy Technology Support Unit (ETSU), *New and Renewable Energy: Prospects in the UK for the 21st Century*, document R-122.

Table 7 Estimated capital and operation costs of gasification projects[13]

Parameter	Arbre demonstration plant costs	Mature technology plant costs
Capacity	8 MWe	33.3 MWe
Start-up date	1998	2010
Equipment and buildings	£18.4 million	£34.1 million
Engineering and development	£5.8 million	£5.1 million
Discount factor	—	12%
Fuel costs	£2/GJ	£1.6/GJ
Electricity conversion factor	35%	44%
Hectares planted	2800	7400
Electricity price	£0.0865/kW	£0.045/kW

these projects is closely related to the waste management policies of the countries in which they are undertaken.

7 Conclusions: Environmental Impacts and Carbon Certification

Emissions of methane to the atmosphere have been rising steadily and in line with population. As it is one of the main greenhouse gases, the control of methane remains one of the top priorities in combating climate change. Since the announcement by Canada and Russia, made at the Johannesburg World Summit, of their intention to ratify the Kyoto protocol, it now seems likely that the three flexible 'mechanisms' proposed at Kyoto will be developed to promote the investment by developed countries in greenhouse gas control projects around the world. Within the Annex 1 countries defined by the Kyoto protocol, the level of emissions of methane from solid waste disposal on land account for around 35% of the total. Figure 5 shows total methane emission for the years 1990–1999 (1990 being the reference year under the Kyoto accord), plotted with methane emissions from landfill. It can be seen that the contribution of methane from landfill to total emissions has barely changed over the decade, despite the increase in the number of landfill gas control projects undertaken in the EU and the impact of the EU 'Landfill Directive' which requires the control of these emissions.

Methane is known to be a stronger greenhouse gas than carbon dioxide, but does eventually oxidize in the atmosphere to form carbon dioxide and water. It is generally accepted that, based on a 100-year scenario, methane has a greenhouse warming potential 21 times that of carbon dioxide. Using this factor, it is conventional to express the effect of the release of methane as a 'carbon dioxide equivalent'. Figure 6 shows the total carbon dioxide emission of the Annex 1 countries compared to the carbon dioxide equivalent of the methane generated from solid waste disposal. This indicates that methane from landfill alone adds around 3.5% of the total emissions of carbon dioxide.

For landfill gas, owing to the greenhouse warming potential of methane, collection and flaring projects provide one of the best environmental investments

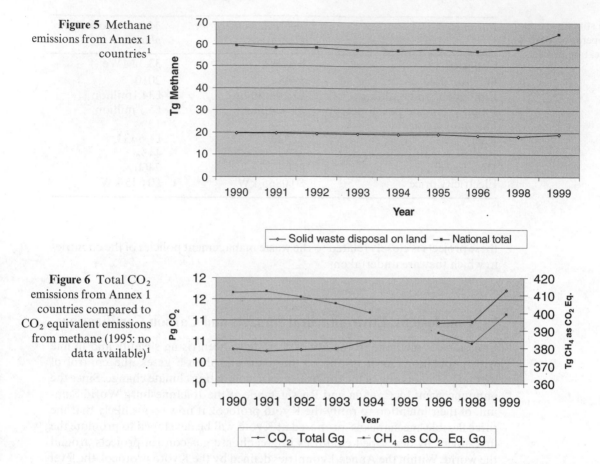

Figure 5 Methane emissions from Annex 1 countries[1]

Figure 6 Total CO_2 emissions from Annex 1 countries compared to CO_2 equivalent emissions from methane (1995: no data available)[1]

in emission reduction technologies. With the addition of power generation, an additional offset of fossil fuel usage can be achieved and preference for power generation over simple flaring should be made.

The environmental value of these projects is slowly becoming realizable, in countries where no support mechanism exists for renewable energy technologies, through the flexible mechanisms provided by the Kyoto accord. The three mechanisms of Emissions Trading, Joint Implementation and the Clean Development Mechanism will eventually put a value on the greenhouse gas emissions saving generated by landfill gas projects and by the fossil carbon offset of biomass projects. Substantial information on these processes can be found at the UNFCCC web site[1] and, although trading will not start in earnest until the first Kyoto period in 2008, there is some early trading in the emission reduction units generated by these projects. With the announcement of the intention to ratify the protocol, made in Johannesburg on 4th September 2002 by Russia and Canada, it now looks likely that the protocol will come into force and, even without the participation of the US, a market for the environmental benefits of methane reduction projects will develop.

Table 8 Contribution of renewable energy to total energy demand in the EU[13]

Type of energy	Actual in 1995		Projected for 2010	
	TWh	% of total	TWh	% of total
Total	2366	100%	2870 (pre-Kyoto)	100%
Wind	4	0.2%	80	2.8%
Total hydro	307	13%	355	12.4%
Photovoltaic	0.03	—	3	0.1%
Biomass	22.5	0.95%	230	8.0%
Geothermal	3.5	0.15%	7	0.2%
Total renewable energy	337	14.3%	675	23.5%

Within the EU, the 1997 report 'Energy for the Future: Renewable Sources of Energy' presents a scenario in which additional capacity will be developed for renewable energy generated from biomass, livestock, sewage and landfill providing 15 Mtoe (million tonnes of oil equivalent), agricultural and forest residues providing 30 Mtoe and energy crops providing 45 Mtoe. Under this scenario the total and projected contribution for renewable energy within the EU is given in Table 8.[14] It can clearly be seen that biomass technologies currently provide, and will continue to provide, the largest single source of renewable energy, with the exception of large-scale hydro plant. Biomass provides the only significant (discounting hydro and the small contribution of geothermal energy) source of base-load power and therefore is a prime area of focus for expansion of renewable energy technology, both in the developed and developing worlds.

While all of the benefits of this form of renewable energy are clear, some attention must be paid to the potential environmental emissions of the combustion processes used in biomass and fugitive methane power-generation technologies. All of these processes do have the potential for the generation of both local and global contaminants in their exhaust streams. The formation of NO_x in high-temperature combustion and its release into the atmosphere has a significant contribution to greenhouse warming and low-NO_x technologies must be developed and utilized for all the processes mentioned.

Biomass technology holds the key to the development of the planned increase in renewable energy resources in the developed world and, by default, provides much of the developing world's local energy demand. Top of the priority list in utilization of biomass must be the capture and use of methane from landfill and other sources, as this has an immediate and significant return in the reduction of greenhouse gas emissions. The utilization of the organic fraction of municipal solid waste in digestion technology requires that this waste is source-separated prior to use as a fuel in order to allow the residues to complete the nutrient cycle as fertilizer. Where more mixed or contaminated organic wastes are concerned, the process of mineralization of the contaminants in pyrolysis and gasification

[14] European Commission, Communication from the Commission, *Energy for the Future: Renewable Sources of Energy*, White Paper for a Community Strategy and Action Plan, COM(97)599 final (26/11/1997).

technologies under controlled conditions offers one of the more promising sources of renewable energy.

In summary, biomass technologies in all of their many forms will provide a sustainable source of energy provided that emissions and soil fertility remain the priority in the design and operation of these plants.

Emissions Trading Schemes: Are They a 'Licence to Pollute'?

FIONA MULLINS

1 Introduction

Objectives

Emissions trading schemes and related mechanisms are becoming the policy instrument of choice for governments and industry around the world. For governments these schemes offer certainty over industry contributions to reducing emissions. For industry they offer a more flexible and cheaper way to comply with government regulations than other policy instruments such as taxes or regulations.

This article focuses on practical experience of the design and implementation of emissions trading schemes. The article examines how companies are responding to the financial incentives that emissions trading provides under a well-known scheme, the U.S. SO_2 trading scheme, and several European greenhouse gas emissions trading schemes that are emerging in response to the Kyoto Protocol: the U.K. pilot emissions trading scheme; the Danish emissions trading scheme; and the forthcoming E.U. CO_2 trading scheme.

Introduction to Emissions Trading

This section briefly explains the broad principles of emissions trading in non-technical terms. The theory of emissions trading is well documented in public economics literature.

Emissions trading is a tool for the efficient redistribution of the cost of achieving an environmental objective. The emission targets that are set give the participants (usually large companies) a right to a limited level of emissions: the targets are quite literally a 'licence to pollute'. Emissions trading reduces the cost of limiting pollution to the aggregate level specified (*i.e.* the sum of all the individual emission targets).

Issues in Environmental Science and Technology, No. 19
Sustainability and Environmental Impact of Renewable Energy Sources

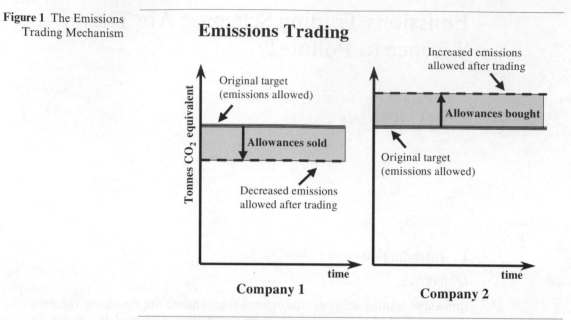

Emissions Trading

Company 1 / Company 2

Source: ERM

Typically, once the targets have been decided, governments give each company a quantity of allowances equivalent to their target. Each allowance represents a quantity of emissions that is 'allowed', for example one CO_2 allowance typically equals 1 tonne of CO_2. Companies are required to monitor and report their emissions and to periodically surrender to the government allowances that are equivalent to their emissions for the period.

Individual companies can buy or sell allowances if they over- or under-achieve their targets. Companies that can reduce emissions at lower cost have an incentive to reduce their emissions more than their target requires. These companies can sell any surplus allowances that they do not need for compliance to companies which find it more expensive to comply with their emission targets, as illustrated in Figure 1. The company targets are amended by purchases or sales of allowances. Thus companies have the flexibility to exceed (*i.e.* not comply with) their original targets as long as they buy an equivalent quantity of emission reductions from another company in the scheme. They will do this if it is cheaper to buy reductions from others rather than to invest in emission reductions in their own operations. Normally (there are exceptions), the total number of allowances in an emissions trading system does not change, regardless of any trading that occurs, and so the environmental objective is assured. The only thing that emissions trading should alter is the cost of meeting the environmental objective.

Emissions trading depends upon a regulatory foundation that limits each company's emissions by law, with tough penalties for non-compliance. Aggregating all of the emission limits (or 'targets') across all of the participants in an emissions trading scheme provides the environmental outcome for the system as a whole. If it works as it is designed to, the fact that individual companies do not restrict their emissions to their target level, and instead purchase emission reductions from others, does not affect the environmental outcome of an emissions trading scheme. This is because reductions in emissions by the other companies compensate for any excess.

This section has outlined how emissions trading works in principle. In practice, emissions trading is rarely implemented from a clean slate, but has to fit in with existing policies. Political and economic constraints can greatly reduce the efficiency of emissions trading. The rest of this article examines what is actually happening in several emissions trading systems that are in place or emerging.[1]

2   Practical experience of emissions trading

Introduction: Emerging Emissions Trading Schemes

One of the first emissions trading schemes, and one of the best known success stories, is the U.S. SO_2 trading scheme established under the 1990 Clean Air Act. The Kyoto Protocol has spurred the emergence of similar schemes for CO_2 emissions in other countries, including Europe (Denmark and the U.K. are the first).

This section reviews the experience of these schemes to illustrate the range of programmes that is already emerging and their economic and environmental effectiveness. Table 1 summarizes some of the main provisions of four schemes that are reviewed in the article, demonstrating the wide range of possible provisions that can apply to such schemes.

U.S. SO_2 Trading

The U.S. SO_2 emissions trading scheme is the forefather of the greenhouse gas emissions trading schemes that are now emerging, several of which are reviewed later in this article. The U.S. scheme was introduced in the U.S. 1990 Clean Air Act amendments and began in 1995 with large coal-fired electric utilities (about 100 plants), and was expanded to all significant utilities (over 2000 plants) in the year 2000. Key features of the program are legally binding emission limits, stringent monitoring and high penalties for non-compliance.

The U.S. scheme set a national goal for SO_2 emissions in the continental United States of 16 million tons[2] of sulfur dioxide (Mt SO_2) per year by the year 2010, a reduction of 40 per cent from 1980 levels. Emissions from the electricity production sector were capped at 8.95 Mt SO_2 per year from 2010 (a reduction of more than 50 per cent from 1980 levels), with annual interim targets stepping down towards this over time. The government allocates annual emission limits

---

[1] In the interests of space and clarity, this article does not address project-trading schemes such as Joint Implementation and the Clean Development Mechanism under the Kyoto Protocol in any detail.

[2] The unit 'tons' is used for the U.S. SO_2 scheme, rather than the unit metric tonnes that is used for the other schemes discussed in this article. One ton is 1.016 tonnes.

Table 1 Summary of selected emissions trading schemes

	U.S. SO$_2$ trading	E.U. (*proposed*) CO$_2$ trading	U.K. GHG trading (*pilot scheme*)	Danish CO$_2$ trading
Legal basis	Legislative (Clean Air Act)	E.U. Directive; Member State legislation; based on existing IPPC permit system.	Contractual agreements between companies and government	Legislative
Gases	SO$_2$	CO$_2$	All six greenhouse gases	CO$_2$
Mandatory or voluntary	Mandatory	Mandatory	Voluntary	Mandatory
Type of targets	Per unit	Absolute (but some leeway for Member States to decide)	Varied: absolute and per unit	Absolute
Compliance	High fines; criminal penalties	High fine (€100; €40 in phase 1)	High future fine of €50 proposed	Low fine
Timing	1993 onwards	2005–2007; 2008–2012	Pilot scheme 2002–2006	2001–2003
Electricity sector participation	Upstream generators	Upstream generators	Downstream energy intensive consumers	Upstream generators

Source: ERM

to each electric utility based on average 1985–1987 energy production levels at standard SO$_2$ emission rates.

The U.S. scheme has proved to be very successful at limiting SO$_2$ emissions which were causing acid rain at a far lower compliance cost than was envisaged before the scheme began. Compared to the cost of traditional regulations, SO$_2$ allowance trading has reduced the costs of reducing acidification. It was estimated in the very early stages of the scheme that the cost of compliance with SO$_2$ trading would be $2.0 billion per year, compared to a cost of $4.9 billion expected from a traditional regulatory approach.[3] In its first phase, with only the largest companies participating, the scheme reduced emissions by millions of tons and the associated acid rain and fine particle pollution reduced by up to 25%. Compliance cost 75% less than originally predicted.[4] Industry estimated initially that the reductions in SO$_2$ would cost them as much as $1500 per ton,[5] and analysts expected the marginal cost of reduction to be between $300 and

[3] GAO *Air Pollution*: *Allowance Trading Offers an Opportunity to Reduce Emissions at Less Cost*, GAO/RCED-95-30, U.S. General Accounting Office, Washington, D.C., December 1994.

[4] B. J. McLean, *Acid Rain Allowance Auction Remarks*, 29 March, 2000, www.epa.gov/ airmarkt/auctions/2000/00remark.html. See also B. J. McLean, *Evolution of Marketable Permits*: *The U.S. Experience with Sulfur Dioxide Allowance Trading*, U.S. Environmental Protection Agency, Washington, D.C., December, 1996.

[5] C. Armstrong, S. Embree and M. Levitsky, *Greenhouse Gas Emissions Reduction Investments Projects in Developing Countries: Identification and Structuring*, International Finance Corporation, June, 2000, p. 3.

$800 per ton, but the price has mostly remained less than or just over $200 per ton.

The large utilities in the first phase of the programme remained well below the overall limits set for them and many were able to meet their targets without needing to purchase SO_2 allowances. Consequently, the price of SO_2 allowances was much lower than expected and many phase one allowances were saved for use in phase two, when the targets got tougher. This was partly due to the fact that the initial allocation of allowances, while requiring an absolute reduction in SO_2, was based on historical production levels ('grandfathering') taking into account the actual SO_2 emitted per unit of energy produced. The targets therefore reflected the starting position of each company, although they did require reduced SO_2 emission rates and an overall reduction in SO_2 emitted. Deregulation of the railways reduced the cost of transporting cheap low sulfur coal and this made the targets easier to meet. More efficient SO_2 scrubbers that remove the SO_2 in the boiler stack came down in price due to increased competition between suppliers and this also reduced the cost of meeting the targets.

The SO_2 allowance is an 'authorization to emit', rather than a 'property right'. This means that the U.S. government could take away utilities' allowances if it needed to. However, participants are certain of their future emission limits. The Clean Air Act legislation specifies the permanent sectoral emission limit and each boiler receives its annual emission limits for 30 years from the time they enter the programme, with a rolling allocation for each year beyond this continuing indefinitely. Utilities continue to receive their annual allocation of allowances each year, even if they close. This means that there is no incentive for utilities to keep old boilers operating just to continue to be allocated allowances and ensures a supply of allowances for new entrants without necessitating any increase in the sectoral emission limit.

The Environmental Protection Agency (EPA) imposes statutory penalties which are very effective at ensuring compliance. Every excess ton of SO_2 over a utility's target incurs a fee which is much higher than the price of SO_2 allowances (the fee was initially set at $2000 per ton SO_2 indexed to inflation, but allowance prices were much lower). In addition to this fee, the EPA deducts one allowance from the following year's allocation for each ton over the target. Company executives can be subject to criminal law if their boilers do not comply.

In summary, the U.S. SO_2 scheme is working to meet the 2010 objective for limiting SO_2 from the electricity sector. The targets were initially seen as tough, which suggests that it would not have been easy to impose the same emission limits on industry without the flexibility of emissions trading. Compliance proved to be far easier than was initially expected by both industry and analysts. The SO_2 trading scheme itself, as well as other events, provided incentives to boost production of sulfur control measures which brought costs down. Giving industry a licence to pollute in the mid-1990s, together with emissions trading, provided incentives to reduce emissions below the targets so that over-compliance has resulted. Over time, the U.S. scheme has proved to be a very effective environmental policy tool that will deliver the desired limits on SO_2 emissions and reduce acidification within the agreed timeframe.

The success of the U.S. SO_2 trading scheme was one factor that led to the

inclusion of international emissions trading provisions in the Kyoto Protocol.[6]
The Kyoto Protocol trading provisions have in turn spawned various regional
and national greenhouse gas emissions trading initiatives, some of which are
discussed below.

Kyoto Protocol

The Kyoto Protocol established emission targets for industrialized countries.[7]
The overall target set by the Protocol is for these countries to reduce emissions of
greenhouse gases (CO_2, CH_4, N_2O, HFCs, PFCs, SF_6) by over 5% below 1990
levels by the commitment period 2008–2012.[8] With the U.S. refusing to ratify the
Protocol and provisions for Russia and others to count additional carbon
sequestration towards their target rather than reducing emissions, the overall
target is now considered to be much less.[9] To achieve this overall reduction, each
industrialized country has a national emissions target which is an 'assigned
amount' based on its agreed percentage reduction (or an increase in some cases)
from 1990 levels.

The Kyoto Protocol and other decisions agreed at the international climate
change negotiations have established a framework for international emissions
trading in order to reduce the costs of achieving the Kyoto targets.[10] The unit of
trade is metric tonnes of CO_2 equivalent.[11] A country that expects not to need its
entire assigned amount to cover actual emissions over the period 2008–2012 can
sell to countries which face high costs to limit emissions to their target level. Any
greenhouse gas permits that are not needed to cover emissions in the period they
were issued will remain valid for future periods until they are used (*i.e.* banking is
allowed from one commitment period to the next). Exactly as in the illustration
in Figure 1, but with countries rather than companies as the traders, international trade in national assigned amounts alters the national targets established
in the Kyoto Protocol against which countries will be assessed for compliance in
2012.

However, emissions trading is not strictly limited to the countries that have
national targets. Additional assigned amounts can be purchased from outside

[6] *The Kyoto Protocol to the United Nations Framework Convention on Climate Change* is the
international climate change treaty agreed in Kyoto, Japan in 1997. Full text is available at
www.unfccc.int.

[7] These are countries listed in Annex I of the Kyoto Protocol and include most OECD countries and
central and eastern European countries that are in the process of transition to a market economy.

[8] Annex I parties' commitments are defined as a basket of six greenhouse gases, which implies that
international emissions trading among governments will include all six gases. A weighting index of
100-year global warming potentials (GWPs) will be used to translate greenhouse gases into CO_2
equivalent units for trading. See Kyoto Protocol text at www.unfccc.int.

[9] See, for example, C. Vrolijk, *Quantifying the Kyoto Protocol Commitments and Mechanisms*,
RECIEL, 2001, **9**, 283–293.

[10] The results of different modelling exercises on the costs of meeting the Kyoto targets with and
without emissions trading are reviewed in J. Weyant (ed.), The costs of the Kyoto Protocol: a
multi-model evaluation, *Energy J. Special Issue*, International Association for Energy Economics,
1999.

[11] A weighting index of 100-year global warming potentials (GWPs) will be used to translate
greenhouse gases into CO_2 equivalent units for trading; see ref. 8.

the group of countries that have emission targets by purchasing emission reductions from developing countries.[12] Thus industrialized countries have a licence to emit a certain level of CO_2 equivalent emissions, with the overall target defined under the Kyoto Protocol (and subsequent decisions). However, the overall level of emissions in developed countries can expand beyond the aggregated Kyoto target level to the extent that certified emission reductions occur in developing countries under the Clean Development Mechanism.[13] The global environmental objective is in theory ensured, but in practice there is potential for certifying emission reductions from developing countries that are not additional to what would have occurred in any case. If this happens, developing country development may be helped by the finance from the industrialized countries, but the environmental objective set at Kyoto will not be met. This type of added complexity, even with the most laudable intentions, increases the chance that the emissions trading that results from the Kyoto Protocol could be a licence to pollute beyond the politically agreed levels.

The present targets for greenhouse gas emissions in industrialized countries for 2008–2012 are not particularly tough for most countries. Some large countries that are in the process of transition to a market economy, such as Russia, have Kyoto targets that give them a national limit on emissions that far exceeds what their actual emissions will reach. Several countries have generous provisions for using sequestration to meet their targets. This is indeed a 'licence to pollute' that is likely to enable global emissions to increase further than they would if Russia were not able to engage in emissions trading and sequestration provisions were tougher. Arguably, however, even the relatively easy targets agreed at Kyoto could not have been achieved without this leniency for Russia, compromise on sequestration and provisions for emissions trading.[14]

If low carbon prices do emerge through international emissions trading, it will be a clear sign that the targets are relatively easy to meet. Achieving stabilization of greenhouse gases in the atmosphere at safe levels will require much tougher targets of more than 60% reduction from 1990 levels over the next 50–100 years.[15] Already informal discussions are beginning on what the targets should be after 2012. The price signal from international emissions trading will help to inform this negotiation.

---

[12] These developing country reductions must be independently validated and certified according to the rules of the Kyoto Protocol's 'Clean Development Mechanism', which aims to provide finance for developing country sustainable production by allowing industrialized countries to use these reductions towards meeting their national emission targets.

[13] The Clean Development Mechanism enables developing countries to attract investment for clean energy and other measures that reduce greenhouse gas emissions. They sell their emission reductions to industrialized countries that have Kyoto targets. The Kyoto Protocol allows industrialized countries to use developing country emission reductions to meet their Kyoto targets.

[14] M. Grubb, C. Vrolijk and D. Brack, *The Kyoto Protocol: A Guide and Assessment*, RIIA/Earthscan, June 1999.

[15] B. Metz, O. Davidson, R. Swart and J. Pan, *Climate Change 2001: Mitigation*, contribution of Working Group III to the Third Assessment Report of the Intergovernmental Panel on Climate Change, IPCC 2001, p. 134. Figure 2.6 illustrates results from a broad range of models showing global CO_2 reductions required for 550 parts per million concentrations of CO_2 in the atmosphere.

Many industrialized country governments will devolve the responsibility for reducing emissions to companies, and companies are likely to be the main participants in both national and international emissions trading if their governments allow this. Examples of the use of emissions trading at the national and regional levels are discussed next.

United Kingdom Emissions Trading Scheme

The U.K. Emissions Trading Scheme (ETS) began in April 2002. It is the most comprehensive emissions trading scheme implemented to date in the coverage of greenhouse gases and industry sectors. The U.K. scheme is complex. The complexity derives from several factors: the scheme's breadth of scope; voluntary company participation rather than mandatory; and interactions with other policies such as an energy tax (the Climate Change Levy or CCL), negotiated agreements (CCLA) with energy intensive industry and renewable energy obligations. There is not space in this article for a full explanation of the U.K. scheme, but it is described in detail elsewhere.[16] This section examines aspects of the scheme that are of most relevance to the question this article is addressing: whether emissions trading is a 'licence to pollute'.

There are several different types of participant in the U.K. ETS, with different methods of allocation, targets defined in different ways, different compliance timeframes and different constraints on trading and on banking allowances for future use. The different participants are discussed below using the terms 'CCLA per unit companies', 'CCLA absolute companies' and 'direct participants' to identify each.[17]

'Direct participants' are the 34 companies that offered to make emission reductions in a government auction in March 2002 in return for an allocation of government money. Any company in the U.K. could participate, with a few exceptions.[18] Government funds were allocated to companies in return for their commitment to annual emission targets to reduce their emissions from an historic base-period (1998–2000) over the 5 years from 2002 to 2006. The direct participants could choose whether to participate and how large a reduction to offer in the auction. They have been paid by the government to take on targets that they have chosen for themselves.

It is likely that the direct participants who entered the U.K. scheme through the auction will be able to meet the targets that they volunteered to adopt. Many

[16] Further information on the U.K. Emissions Trading Scheme is available on the government website: www.defra.gov.uk/environment/climatechange/trading/index.htm; see also chapter 6 in this book.

[17] Climate Change Levy Agreement (CCLA) companies are those that have agreed to limit their energy or CO_2 in return for an 80% rebate on the Climate Change Levy (CCL). The CCL is charged on all industry and business use of energy (electricity, coal and gas), and is revenue neutral in that the money raised is recycled *via* reduced employer taxes (national insurance). CCLA companies tend to be those with high energy intensity and relatively few employees that would therefore be relatively worse off under the CCL than companies with low energy costs and many employees.

[18] Electricity generators and transport were excluded as it is envisaged that their future participation will be through other mechanisms that are not yet defined.

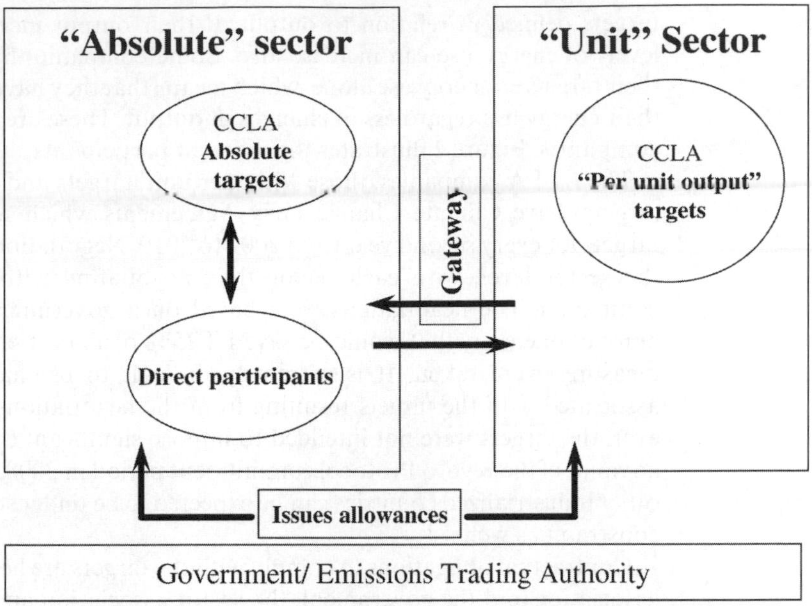

Figure 2 Participants in
the UK Emissions
Trading Scheme
(CCLA = Climate Change
Levy Agreement)

Source: ERM

of these companies are likely to be those with stable or decreasing emission trends so that reductions from an historic base-period will not threaten their economic growth prospects. Although there are only 34 direct participants, some of them have very large annual emissions, and so they are a significant component of the U.K. scheme. It is also quite possible that some of these participants have targets that represent an allowance for higher emission levels than their actual emissions will reach. For example, if a direct participant's output reduces dramatically, reducing the associated emissions, their targets will remain the same, giving them surplus allowances that they will be able to sell on the U.K. market. If output increases dramatically, they could face shortfalls and have to purchase on the U.K. market.

The direct participants have targets defined as tonnes of CO_2 equivalent and fairly strong incentives to comply, as they lose all of their government incentive money if they do not comply and their future allowance allocations are reduced. In the future, the government plans to legislate a penalty fee of £30 per tonne CO_2 for all U.K. scheme participants in line with the proposed E.U. emissions trading penalties, which will provide legislative backing for compliance penalties rather than relying on loss of incentives alone.

The majority of participants in the U.K. scheme (over 6000 entities) are companies in energy intensive sectors that have negotiated targets for limiting their energy use in return for a partial exemption from the Climate Change Levy. Most of these companies' targets are defined as limits on their energy use per unit of output over 12-month periods that vary slightly from sector to sector and are based on different base years. These are the CCLA 'per unit' companies with targets defined in relation to output. If their output increases, their absolute levels of energy use can increase also. Some companies have chosen to define their targets as energy use alone, which means that they have an absolute limit on their energy use regardless of changes in output. These are the CCLA 'absolute' companies. Figure 2 illustrates the different participants.

The CCLA companies (those with 'per unit' targets and those with 'absolute' targets) have Climate Change Levy Agreements which set negotiated energy targets for every second year from 2002 to 2010. Negotiations were carried out at the sector level, with each sector then negotiating with individual member companies. The negotiations were based on a government assessment of the amount of energy that would be saved if 75% of all cost-effective energy saving measures were taken. It is extremely difficult to estimate compliance costs associated with the targets resulting from the negotiations at this point. However, the targets were not intended to impose significant costs on companies in advance of the Kyoto Protocol commitment period in 2008, when competitors in other industrialized countries can be expected to be under some sort of emissions constraint as well.

Contractual obligations to comply with the targets are held between the sector association and the government, the sector association and each member company, and each company and the government. In return for complying with their targets, the companies receive 80% exemption from the CCL for the following two years. If they do not comply they must pay the full CCL for the following two years. Full CCL payment is a large cost for energy intensive companies and this is a powerful incentive for compliance.

Compliance can be achieved by a sector as a whole, in which case individual company compliance is not checked. If a sector fails to comply, then individual member companies must comply in order to receive the CCL exemption. The CCLA companies may use CO_2 allowances purchased from other participants in the U.K. ETS for compliance if their energy use exceeds their targets. Tonnes of CO_2 are converted to units of energy on the basis of the fuel used by the company. Once the companies have demonstrated that their energy use per unit of output (for 'per unit' targets) or their absolute energy consumption (for 'absolute' targets) is less than the target, the government places CO_2 allowances in their company account in the registry that are equivalent to the amount of energy saved. Thus the CCLA companies are able to participate in emissions trading and, if they take up this opportunity, will form the majority of entities in the U.K. ETS.

The CCLA companies only receive their allocation of allowances after they have their energy and output levels for the milestone period verified and demonstrate that their energy use is below target. Following this, the government will allocate CO_2 allowances equivalent to the additional energy savings that they have made. The fact that (as at September 2002) the companies did not yet have their allowances was one factor that limited the supply of allowances on the ETS market and kept allowance prices high. Prices for allowances rise steadily from £4.50 per tonne at the beginning of the scheme to over £12 per tonne in October 2002 and then fall below £4.00.

Another factor contributing to higher than expected prices was that the 2002

CCLA targets for some sectors turned out to be more difficult to meet than many expected. Lower than expected output does not lead to equivalent decreases in energy use in many industries, so the global reduction in demand and strong pound made the targets more difficult than they would have been if output were stable or growing. Thus for 2002 at least, the CCLA targets appear to be effectively limiting energy use and therefore CO_2 emissions, and the price of carbon was correspondingly high.

Because many CCLA participants have per unit targets, a 'gateway' has been established between the 'absolute' and the 'per unit' sectors to avoid inflation of allowances available to the 'absolute' sector in periods of output growth. The gateway will close whenever aggregate sales from the 'per unit' sector to the 'absolute' sector equal the sales in the other direction. Effectively, trading will only take place between the two sectors when the marginal cost of abatement is lower in the 'absolute' sector than in the rate-based sector. The gateway prevents growth in the 'per unit' sector of the ETS from inflating the allowances available to participants with absolute targets.

The U.K. is gaining valuable experience with its pilot scheme. The initial targets are not particularly difficult for most participants to meet, but the allowances on the market are selling at a robust price: a lot of emission reductions could be achieved in the U.K. at less than £12 per tonne of CO_2. As the market matures, the price will be one good indicator of how tough the targets are. As soon as the E.U. Directive is agreed, the U.K. will have to make the transition to E.U. trading. Many aspects of the U.K. scheme will be changed when they implement the Directive, including which industry sectors future targets are placed on, how the targets are defined and how tough the targets are. The experience of the U.K. pilot scheme will inform this transition.

Danish Emissions Trading Scheme

The Danish Act on CO_2 Quotas for Electricity Production entered into force on 15 July 2000 with effect from 2001, launching the first CO_2 emissions trading scheme. The scheme is limited to large electricity generators operating in Denmark (including two foreign-owned companies) and is scheduled to end in December 2003. These companies export carbon intensive electricity to Norway and Sweden in dry years, causing Danish emissions to increase dramatically compared to years when Denmark imports hydro electricity. The company CO_2 quotas and emissions trading scheme are designed to cap the amount of CO_2 associated with electricity production so as to ensure that Denmark can meet its international environmental commitments under the Kyoto Protocol.

The Danish system is based on an annual, national ceiling for the allowable emissions from electricity production. It is limited to the electricity sector, responsible for about 40% of the total emissions of CO_2 in Denmark. The ceiling is reduced each year, from 22 million tonnes in 2001, to 21 million tonnes in 2002 and 20 million tonnes in 2003 when the scheme ends.

The Danish system limits total CO_2 emissions from all Danish power production (including around 500 power producers) for the years 2001–2003. The emissions trading component focuses solely on the eight largest electricity producers. These eight very large power companies are given an allocation of allowances of about 70% of the level of emissions in the base period 1994–1998. The companies are allowed to trade their allowances and to save (or 'bank') allowances for use in future years. The companies must surrender allowances equal to their emissions each year. The penalty for failing to hold sufficient allowances is 40 Danish kroner (about €5.40) per metric tonne of CO_2. Smaller Danish electricity companies are not included in the emissions trading aspect of the scheme. Total CO_2 emissions by the electricity sector will fall from 23 million tonnes in 2000 to 20 million tonnes in 2003, compared to an average of just over 30 million tonnes annually during the period 1994–1998.[19]

The Danish penalty payment of €5.40 per tonne of CO_2 effectively caps the allowance price at a low level compared to model estimates for world carbon prices. In the first trading period of 2001, seven allowance trades were carried out, trading a total volume of 260 000 tonnes of CO_2 at an average price lower than the €5.40 non-compliance penalty tax.[20] No company paid the penalty tax, so it appears that the allocations of allowances in Denmark for 2001 capped the emissions at a level which could be met from within the Danish system without having to implement emission reduction measures at or above the penalty level. This could be because the supply of hydroelectricity was at a reasonable level since the scheme began in 2001 compared to the historical 1994–1998 period so that the electricity companies were not generating enough coal-fired electricity to exceed their allocations. At least five swaps of certain amounts of Danish allowances for verified emission reductions under other systems were carried out in 2002, but no information is available on the details of these exchanges other than that the Danish allowances were not swapped one-for-one with other verified emission reductions.

From the information available to date, it is clear that the Danish 'licence to pollute' has not been a serious limitation on electricity producers. However, the Danish scheme is certainly having the effect of placing a cost on CO_2 which is reflected in the demand for hydroelectricity. It is possible that Danish companies limited their exports to the level at which they could remain within their allocations, with the balance of energy supply imported. For Danish companies to produce above this level, the Norwegian or Swedish importers would have had to be prepared to pay the Danish penalty rate on top of the cost of the electricity. In a dry year this might happen, passing on the cost of increased Danish CO_2 emissions to the buyers. If so, the Danish penalty rate would not provide sufficient incentive to ensure that Danish Kyoto commitments are achieved.[21] The Danish government will consider whether and how to extend

---

[19] *Danish Environment & Energy Newsletter*, October 2000 (www.mex.dk/uk/), Danish Energy Agency.

[20] Natsource reported a Danish bid/offer spread of U.S.$2.14–2.85 per metric tonne of CO_2 as at March 2002, *Assessment of Private Sector Anticipatory Response to Greenhouse Gas Market Development: Final Analysis*, July 2002, Natsource LLC with GCSI, available from www.natsource.com.

[21] Danish CO_2 cap and trade scheme update, February 2002; S. L. Pedersen, Ministry of Economic and Business Affairs http://www.ens.dk/graphics/ENS_Forsyning/Kvoter/DK_ETR_update190202.doc and in RECIEL, 2000, 9. More detail can also be found on the DEA website: www.ens.dk/uk/energy_reform/emissions_trading/index.htm.

the emissions trading scheme after 2003 and intends to join the E.U. scheme when it is implemented, which would establish a much higher penalty rate of €40 per tonne of CO_2, increasing to €100 from 2008.

European Union

The European Commission has proposed a Directive that would establish an E.U.-wide emissions trading program for specified sources of CO_2 emissions. Existing schemes such as the Danish scheme (if it continues beyond 2003) and the U.K. scheme will be brought into line with the final E.U. scheme. The Directive is in the process of being debated in the European Parliament and Council. The final Directive is expected to be adopted in late 2003. The first phase of the programme is scheduled to begin in 2005, with a second phase from 2008 to 2012.

Participation is intended to be mandatory for about 5000 installations in specified energy-intensive sectors such as energy combustion, refineries, iron and steel, cement and pulp and paper. The total quantity of allowances issued and their distribution to participants is left to the Member States. Each Member State must submit a national allocation plan in advance to the Commission. The quantity issued must be in line with Kyoto commitments, energy and climate policy and the reductions that are achievable.

The Directive specifies that allowances should be distributed free during the 2005–2007 period. However, there has been an attempt to introduce an amendment requiring a certain portion to be auctioned, and the allocation method from 2008 may require auctioning. If allowances are auctioned, then the companies will purchase their initial allocation. The price that the allowances sell for at auction will indicate the abatement cost across the trading system. However, apart from the initial investment in purchasing the allocation, companies should in theory behave in precisely the same way in subsequent trading, valuing their allowances at their own cost of reducing emissions or at the market price.

In the initial period, when allowances are distributed free to the companies that are required to participate, governments must decide on the level of allocation to give each company. Each country must produce a national allocation plan with objective and transparent criteria for the distribution of allowances. The distribution of allowances to participants must be consistent with the E.U. requirements regarding state aid and must treat new entrants fairly.

As in the U.K. scheme, each allowance entitles the holder to emit greenhouse gases equal to one tonne of CO_2 equivalent. Member States will have the option to allow extra tonnes saved during the 2005–2007 period to be carried forward into the 2008–2012 period. Companies will be required to surrender allowances equal to their actual emissions during the previous calendar year. The penalty for non-compliance will be loss of allowances equal to the excess emissions plus a financial penalty equal to €100 per metric tonne of excess emissions. During the 2005–2007 period the financial penalty will be €40.

It is much too early to comment on the efficacy of the E.U. scheme, given that its final design and provisions are not yet agreed. It is very likely that individual Member States will decide or negotiate targets with companies, although the allocations that are decided at the Member State level will be checked at the

European level to ensure that the targets are not causing competitive distortions within the European Community. In any case, any leniency on emission traders will increase the burden of meeting the Kyoto targets on other sectors that are not involved in emissions trading, such as transport and households. This type of transfer of the climate burden would not be politically feasible. In countries that are committed by international treaty to limiting national emissions, it is unlikely that their industry will be given emissions trading targets that are a 'licence to pollute' beyond the levels that are necessary to achieve the national Kyoto targets.

3 Conclusions

A key feature of emissions trading schemes is that the cost savings that emissions trading brings enable greater emission reductions than would otherwise be the case. The cost savings make tougher targets more acceptable than they would be without trading. Some observers have commented that the national greenhouse gas emission limits (or 'assigned amount') that were negotiated in Kyoto in 1997 could not have been agreed without the inclusion of emissions trading provisions in the Kyoto Protocol.

There are differences in the political will of governments to impose environmental costs on industry, and differences in the costs imposed on different sectors. For example, the fuel and energy taxes implemented by different governments suggest a very wide range of implied costs of carbon in different countries. Table 2 illustrates a few of these. With emissions trading, however, better information on the cost of reducing carbon in different countries and worldwide will emerge.

Costing studies of schemes that have been implemented show that cost estimates for meeting emission limits are often much higher before implementation of an emissions trading scheme than after (*e.g.* SO_2, CFCs). This is partly due to events that are not to do with emissions trading (such as railways deregulation in the U.S. reducing the coal transport costs and so reducing the SO_2 trading scheme compliance cost), but partly also to the impact of the emissions trading mechanism itself.

Rather than having an environmental manager ensuring that the company meets its obligations at least cost, with emissions trading, operational and financial managers are made aware of the financial implications of greenhouse gases and have an incentive to seek out innovative low-cost reduction opportunities. Company traders (*e.g.* in energy companies) and brokers help companies to arbitrage between the opportunities. Companies may not know how much it will cost to reduce greenhouse gas emissions. These gases have not been regulated before and awareness of opportunities to reduce emissions is only now becoming commonplace in day-to-day operations of some of the largest multinational companies. Under emissions trading, a market value for carbon gradually is established that reflects the lowest possible cost for meeting the aggregate emissions limit across all participants in the scheme. While the 'growth in greenhouse gas awareness' effect is not captured in models, it is evident from the

Table 2 Implied cost of carbon from regional and domestic policies

Jurisdiction	Implied cost $\text{€2001}/tCO_2$	Policy
Austria	5.13	Electricity levy
Denmark	0.32	Carbon tax on electricity
Italy	30.08	Carbon tax
Netherlands	1.35	Energy tax
UK	320.00	Climate Change Levy on electricity, gas, coal

Source : ERM review 2001

trading schemes that have been implemented that the incentives work differently for trading schemes than, for example, for taxation.

Emissions trading is not the right tool for every pollutant. Emissions trading works well only where it does not matter where the emission reductions occur (no local effects), where it is mandatory and the regulatory backing of the scheme limits emissions to an environmentally beneficial level, and where political and existing policy issues do not require too many exceptions and complexities which reduce transparency and increase costs. In practice, emissions trading works best to address the emissions from large firms, owing to the administrative require-ments for verification, monitoring and reporting that are too costly for small firms and to reduce the administrative burden on government of compliance checking.

The emergence of this new policy tool is dynamic. Emissions trading interacts with other policies and evolves over time with growth in experience and industry acceptance of the limits around which trading occurs and the emergence of market mechanisms that facilitate market liquidity and depth.

Emissions trading can be a 'licence to pollute' without constraining companies to reduce emissions from business as usual in the early stages, but is likely to be an increasingly effective tool for addressing greenhouse gases and other pollu-tants in the longer term.

UK Government Policy on Renewable Energy

BRIAN WILSON

1 Introduction

1 April 2002 saw a significant development in UK Renewables policy: the introduction of the Renewables Obligation on licensed electricity suppliers. This is a key part of the comprehensive framework which we have put in place to help us to meet our target of securing 10% of electricity sales from eligible renewable sources by 2010. In essence, it will establish a growing demand for renewables that will give developers and their financiers the necessary assurance to make long-term investment decisions.

The Renewables Obligation on electricity suppliers in England and Wales, and the equivalent Scottish Renewables Obligation in Scotland, require electricity suppliers to provide a rising proportion of their sales from eligible renewable sources until 2010, from 3% in the first period to 10.4% by 2010/2011. This effectively calls for a six-fold increase in our portfolio of eligible renewables. This rate of growth is challenging but certainly achievable.

The growing focus on renewables is a key element in our overall programme for tackling climate change, and in our energy policy objectives of ensuring greater sustainability and diversity in the overall energy mix. Renewable energy has a crucial role to play in contributing to our Kyoto target of reducing carbon dioxide emissions to 12.5% below 1990 levels by 2010. The Obligation, designed to stimulate the industry and enable us to meet (and indeed exceed) our European indicative targets, will remain in place until 2027 and be supported by an extensive package of funding to bring forward a range of technologies.

The Obligation's introduction marked the culmination of a three-year period of consultation and close contact with the industry, where the future of renewable and sustainable energy policy was shaped and refined. However, this cannot and should not be the end of the story. The recent Performance and Innovation Unit (PIU) Energy Review,[1] instigated by the Prime Minister, shows that renewables are high on the political and social agenda. The recently published

[1] Performance and Innovation Unit, *The Energy Review*, London, 2002.

Issues in Environmental Science and Technology, No. 19
Sustainability and Environmental Impact of Renewable Energy Sources
© The Royal Society of Chemistry, 2003

Energy White Paper, also underlines the importance of a secure, sustainable and diverse energy policy in which renewables will make a growing contribution in the years to come. This rising profile is certainly reflected in the increasingly large volume of correspondence that I receive on a daily basis, ranging from concerned schoolchildren to individuals wishing to know how they can do more to play their part in driving renewables forward.

The breadth of activity now and the focus of our policy has certainly shifted since our earliest attempts to promote renewable energy.

2   NFFO and Initial Success

The collection of renewable energy statistics effectively began in 1989 when all relevant renewable energy sources were identified and, where possible, information was collected on the amounts of energy derived from each source. This enabled the development of a two-part strategy: the Non-Fossil Fuel Obligation (NFFO) and a complementary support programme. That initial programme, executed in partnership with the industry and the European Commission, enabled assessment of the resource available, development of near-market and medium-term technologies, and stimulation of the industry through a relatively generous funding package.

As the first major instrument for pushing forward the development of renewables capacity, NFFO [including NI-NFFO (Northern Ireland Non-Fossil Fuel Obligation) and the Scottish Renewable Obligation (SRO)] aimed to assist the renewables industry by allowing premium prices to be paid for electricity for a fixed period. We were very successful in introducing renewables to the market, and in bringing costs down to a more acceptable level. As of 30 June 2002, 402 projects were operational, representing a generating capacity of 1003 MW declared net capacity.

However, as the structure of energy markets began to change—with the separation of the supply and distribution channels, and the impending introduction of the New Electricity Trading Arrangements (NETA)—it became clear within Government that this represented an opportunity to consider more innovative and market-focused options for support. In response to this, my predecessor John Battle published *New & Renewable Energy: Prospects for the 21st Century*[2] to explore the feasibility of a range of support mechanisms, and invite industry comment upon them.

3   Towards the Obligation

Subsequently, in February 2000, the Government set out the conclusions from this initial consultation exercise.[3] Work then started in earnest on developing what the industry had considered as the most appropriate form of Obligation: supplier based. Unlike NFFO, this new system would not be based on a system

[2] Department of Trade & Industry, *New & Renewable Energy: Prospects for the 21st Century*, London, 1999.
[3] Department of Trade & Industry, *New & Renewable Energy: Prospects for the 21st Century: Conclusions in Response to the Public Consultation*, London, 1999.

of long-term premium price contracts with individual generators. NFFO had also focused on technology bands, and again this would not be replicated in the new mechanism.

Instead, it was decided that the new Obligation would call on individual licensed electricity suppliers to prove that they had supplied a certain percentage of their sales from renewable-sourced electricity. That percentage would rise steadily to 10% in 2010.[4]* Compliance would be assessed by Ofgem (the Gas & Electricity Markets Authority) using a system of certificates, now known as ROCs (Renewables Obligation Certificates). ROCs would be issued to accredited generators, and be redeemed by suppliers as evidence that they had supplied renewable-sourced electricity to a customer in the UK. To encourage dynamism, choice and trade within the market, it was agreed that ROCs could be sold independently of the electricity with which they were associated. This would then mean that a ROC's definition could be widened to provide evidence that another party had supplied renewable-sourced electricity on a particular supplier's behalf.

As the detail of the Obligation was refined, again in close consultation with the industry and Ofgem, work was also ongoing on developing the other planks of our policy: exemption of renewables heat and electricity from the Climate Change Levy (introduced in April 2001); refining our existing R&D support programme; and looking at ways in which renewables could be supported at the regional level.

In developing the Obligation, a key caveat was that any additional cost should be set at an acceptable level for the consumer. For too many families and individuals in the country, fuel poverty continues to be a very real problem and we were simply unwilling to let our environmental objectives override their particular needs. By the same token, we were reluctant to create excessive additional costs for the industrial and commercial users. These concerns were the impetus for introducing the Buy Out Price mechanism. While some initially argued that the Buy Out would provide an 'easy way out' for those suppliers with a disinclination to support renewables, the concept of recycling those payments to competitors was introduced as a further incentive to supply.

In October 2000, the then Secretary of State launched our preliminary consultation[4] exercise—appropriately at the Greenpeace Business Conference, showing that there can and should be a synergy between environmental considerations and big business. As he spoke, the importance of what we are trying to achieve was brought home by severe floods throughout the country—evidence of climate change in action.

As responses to the consultation paper trickled then poured in, we continued

---

* Large-scale hydro (initially over 10 MW) and energy from waste were both originally considered to be key to helping us meet our overall 10% target. However, it was felt that both were sufficiently well-established not to need the additional support that would be provided by the Obligation. Given this, the Obligation target was calculated at this stage as 7.7% of total electricity supply in 2010. This decision was superseded by the definitions of eligibility set out in the later Statutory Consultation exercise.

4 Department of Trade & Industry, *New & Renewable Energy: Prospects for the 21st Century: The Renewables Obligation Preliminary Consultation*, London, 2000.

to develop the policy. Questions of cost and eligibility were explored further, and it became clear from the responses that opinion was divided on issues such as the role of energy from waste, while others questioned proposals for borrowing ROCs.

We also introduced our proposals for capital grants to push forward the development of offshore wind and energy crops. As more sources of funding became available, we were able to think more creatively about our supporting mechanisms.

The Utilities Act 2000 required us to consult widely with the industry and relevant consumer groups before our proposals could be laid before Parliament. Accordingly, our final detailed proposals were set in our Statutory Consultation document[5] in August 2001. In April 2002, the new Renewables Obligation and analogous Renewables Scotland Obligation came into effect. The target proportion for renewables for the first year is 3%, rising to 10% in 2010. Eligible sources include all renewables, but there are some very specific exclusions, namely:

- Existing hydro plant of over 20 MW
- All plant using renewable sources built before 1990 (unless refurbished)
- Energy from mixed waste combustion unless the waste is first converted to fuel using advanced conversion technologies

Only the biodegradable fraction of any waste is eligible, in line with EU Directive 2001/77/EC on the promotion of electricity from renewable energy. All stations outside the UK (including its territorial waters and the continental shelf) are also excluded.

4 Supporting the Obligation

The Obligation is underpinned by a strong package of support mechanisms, both financial and non-financial. Taken together, these are encouraging the industry to grow, while simultaneously removing some of the barriers to that growth.

Financial Incentives

At least £260 million is available to help industry develop the necessary capacity to meet our targets. At present, that package is broken down as shown in Table 1. This package stems from a number of sources, including: my own department; the PIU; the New Opportunities Fund; and the Department for the Environment, Food and Rural Affairs (DEFRA), who are providing establishment grants for energy crops worth £29 million over a six-year period. This will complement our own Bio-Energy capital grants scheme, which aims to promote the efficient use of biomass for energy and, in particular, the use of energy crops by stimulating the early deployment of biomass fuelled heat and electricity generation projects. It will do this by awarding capital grants towards the cost of equipment

[5] Department of Trade & Industry, *New & Renewable Energy: Prospects for the 21st Century: The Renewables Obligation Statutory Consultation*, London, 2001.

Table 1 Breakdown of government funding for renewables: 2001–2004

Item	Budget (£ million)
Offshore wind	74.0
Energy crops	95.5
Photovoltaic (PV)	20.0
Biomass heat	3.0
Community projects: Clear Skies programme	10.0
Wave and tidal	5.0
Advanced metering and control	4.0
DTI New & Renewable Energy R&D programme	55.5
'Blue skies research'	10.0

in complete working installations. The scheme has a number of priority areas, which should enable over 100 MW of plant to be commissioned. These include:

(1a) Large-scale electricity generation, greater than 20 MWe (megawatt equivalent) maximum rated output.

(1b) Medium-scale electricity generating, or Combined Heat and Power (CHP), installations over 1 MWe output with a preference, but not a requirement, for CHP with high overall efficiencies.

(2) Demonstration of new, high-efficiency technology for generating electricity from energy crops.

We are equally keen to see offshore wind fully develop its potential and the capital grants programme has been specifically developed to kickstart development and to stimulate early deployment of this important resource. I was naturally delighted to see so many developers benefiting from the Crown Estate's preliminary granting of sea-bed leases. Nineteen such developments have been put forward. The first of these projects, Scroby Sands, North Hoyle, and Rhyl flats have already received the green light to proceed with development.

We have already had some success in this country with wave and tidal power, and the £5 million shown in Table 1 will be channelled through the New and Renewable Energy R&D programme. The world's first commercial wave power device on Islay was supported by the Department of Trade and Industry (DTI), and we are continuing to support Wavegen, the company involved, as it refines and further develops the technology involved. Last May, I announced funding of up to £2.3 million to support their development and demonstration of a series of new wave energy devices off the Western Isles.

A further £1.6M will go to Tidal Hydraulic Generators Ltd (THGL) for development and testing of their tidal stream prototype. We are determined to ensure that the UK remains at the forefront of this technology. We are already providing financial backing, through our New & Renewable R&D programme, to the Engineering Business Ltd. This £1.8 million project will install a 150 kW tidal stream demonstration device, known as Stingray, in Yell Sound in the Shetland Islands. My department is providing 75% (£1.1 million) of the funding to design, build, install, operate, decommission and evaluate the results from the Stingray demonstrator. The remainder of the £1.8 million cost will be provided by The Engineering Business. Initial estimates of unit energy costs suggest that Stingray will be competitive with other renewable energy technologies. Given this, it is my hope that many more projects like this will come on stream in future.

Planning

The allocation of funding also points to another major plank in our policy: a regional strategic approach to planning for renewables. I recognize that, in several cases, the planning process has not helped our drive to increase deployment: we are trying to overcome this problem in a number of ways. We have already introduced locational flexibility for NFFO projects; this means that those developers who have failed to secure permission to develop their chosen site now have the option of finding another, more suitable, location.

In addition, each of the regions completed an assessment of the potential for renewables in their own area. This will enable them to move a step further by focusing on those technologies with most local potential, and improving the profile of renewables within the local planning community. We have already allocated £1.7 million of the £2.5 million which is available to help the regions achieve their goals, and to spread best practice within the planning community.

The Office of the Deputy Prime Minister (ODPM) is also committed to revising the planning guidelines on renewable energy—PPG 22. Scotland has already led the way in this, and has produced revised guidelines for developers and local authorities which stress the importance of renewables to overall Government policy on climate change.

That does not mean that we want the countryside to be covered with wind turbines or any other type of renewable project but of course new construction will occur. What I see as a vital ingredient for positive planning will be the question of balance and sensitivity. I believe that proposals must be considered with an open mind. It is remarkable what can be achieved through sensitive design and architecture. We are not looking to impose these developments and we expect developers to work in sympathy with the local environment.

Renewables UK

In March 2002, I launched the new Renewables UK unit in Aberdeen. Its overall objective is to maximize the UK's involvement in renewables projects both at home and abroad. We decided that the unit should be located in Aberdeen because of the proximity of major players involved in the oil and gas industry, who are ideally placed to take advantage of diversification opportunities. It is also recognized that the UK renewables industry is in a comparable state of development to that of the oil and gas supplies industry 30 years ago. Co-location within the Department's Oil & Gas Industry Development Unit will also enable it to take advantage of the extensive bank of knowledge and expertise that exists.

The Unit's detailed immediate objectives for the coming year are:

• An offshore renewables study to identify key markets and future technologies

- Establishment of a long-term business support strategy for wave and tidal
- A UK renewables gap analysis to identify key strengths and areas for development
- Developing relationships with the industry

Renewables and The Smaller Generator

We have been working equally hard to put together the right framework to ensure that electricity transmission and distribution issues do not present technical and economic barriers to the future development of renewables and embedded generation—generation capacity which tends to be small-scale, localized and non-grid connected. If we are to meet our targets, then such forms of generation are likely to increase in importance with time. The DTI/Ofgem-led Distributed Generation Working Group is monitoring progress on the implementation of a wide range of measures to facilitate easier connection to the distribution networks.

Ofgem published their report on the first year of operation of the New Electricity Trading Arrangements (NETA) on 24 July 2002. Previously, concerns had been expressed that renewables would be in a disadvantageous position under NETA. However, the smaller generators who responded to Ofgem's survey reported that they produced the same amount of electricity in the first year after the introduction of NETA as they had in the year before. The first-year report also notes that Ofgem has changed the balancing and settlement code to help intermittent generators, and has issued proposals to encourage the growth of independent consolidation services.

Smaller generators will also be encouraged through my department's Clear Skies scheme. This is worth £10 million and supports small-scale projects across a variety of technologies. We have already provided £800 000 for a similar scheme: the Community Renewables Initiative. Administered by the Countryside Agency, this particular scheme provides support for community-based initiatives in 10 distinct English regions.

5 What Next for Renewables?

The recent PIU Energy Review included an examination of the longer term prospects for renewables. PIU suggested that the focus of our future policy should be to establish new sources of energy that are, or can be, low cost and low carbon. We consulted widely on the Review and this has influenced our future policy which has been set out in the recently published Energy White Paper.

Obviously, I want to see renewables grow in stature and importance, and we have to ensure that any targets we set are achievable ones. The foundations are now being put in place to ensure that this growth can be achieved. In our quest to nurture, strengthen and expand the growth of renewable energy technologies, we are definitely making a positive difference—for the environment, for our future energy needs and towards meeting the targets laid down in the Kyoto Protocol. The challenge now is for industry and consumers to embrace the technologies and help us to move forward.

Given our efforts on the domestic level, we were naturally disappointed that the recent World Summit on Sustainable Development in Johannesburg failed to adopt specific targets for renewable energy. However, the European Union, whose view we obviously share, has expressed its strong commitment to the promotion of renewable energy and increasing the share of renewables in the global total primary energy supply. We recognize that increasing the use of renewables is an essential element of achieving sustainable development at both national and global levels.

As I indicated above, our own domestic targets are challenging but achievable. The Renewables Obligation has given a significant boost to investor confidence in this new emerging market. Appropriate investment in this area will also create significant business opportunities which cannot be ignored, both for the UK market and the environmental export market. I am keen to see UK companies taking the lead in this field, as well as contributing to our aim of a 60% reduction in CO_2 emmissions by 2050. That would indeed be the icing on the cake.

Renewables, Sustainability and Precaution: Beyond Environmental Cost–Benefit and Risk Analysis

ANDREW STIRLING

1 Assessing the Sustainability of Energy Options

Energy systems have long lain at centre stage in the global sustainability debate. Recent decades have seen the intensification of acute concerns over issues such as greenhouse gas, acid and particulate emissions, nuclear waste and accident risks and (in some quarters) the ecological and landscape effects of large-scale deployments of renewable energy technologies. Although there is sometimes a tendency to take 'climate change' as a proxy for energy sustainability considerations, the full range of issues are in fact highly diverse. The implications for practical decision-making over choices between contending energy options are often far from straightforward. Obvious examples concern the continuing vexed debate over the relative sustainability merits of nuclear power and renewable energy and the relative priority to assign to development of different renewable energy technologies.

A pressing need therefore arises for robust ways to inform policies aimed at the pursuit of sustainable energy strategies. In the energy sector, as elsewhere, there currently exists a state of considerable ambiguity and tension around what are held to be two quite distinct approaches to regulatory appraisal. Should the assessment of the relative sustainability of different energy options be based exclusively on particular well-known scientific and technical considerations, or should it be more 'precautionary'—paying greater attention to scientific uncertainties and social and cultural factors? Although this paper will argue that the distinction is misleading, this dichotomy between 'scientific' and 'precautionary' approaches is a high-profile feature of general debates on sustainability—and the energy sector is no exception.

A 'scientific' approach to the assessment of energy sustainability is—as in other sectors—conventionally taken to imply reliance on quantitative tech-

Issues in Environmental Science and Technology, No. 19
Sustainability and Environmental Impact of Renewable Energy Sources
© The Royal Society of Chemistry, 2003

niques. These are used to aggregate across a range of those different types of effect that are held to be measurable. These are then taken as a proxy for a wider array of issues that are less readily quantified, whose relevance is contested, or which remain uncertain. Key examples of this type of approach may be found in comparative risk assessment and environmental cost–benefit analysis. For reasons that will become clear later in this article, these can be described as '*risk-based*' techniques. Such methods have been employed extensively in the regulation of the electricity supply industry. Indeed, they have been developed in this field to a state of elaboration and maturity matched in few other sectors. The basic aspiration underlying the use of these 'risk-based' techniques is that—in and of themselves—they offer a robust means to prescribe and justify regulatory decision-making. The high financial and political stakes attending discussions of global sustainability make the need for policy justification especially crucial. To this end, the authority of 'risk-based' approaches rests largely in a general appeal to scientific rigour and in the apparently clear prescriptive nature of the analytical results thereby obtained.

For its part, a '*precautionary*' approach reflects a rather different perspective, introducing a wider range of emerging issues in the general sustainability debate. At root, a precautionary approach contrasts with the more reductive 'risk-based' approach in extending equal attention to those effects that may be less readily quantifiable. It addresses themes such as complexity, variability and the potential for non-linear vulnerabilities in natural and social systems. It highlights the consequent potential for 'surprises' affecting all manner of options. Precaution places greater emphasis on active and dynamic choices between technology and policy alternatives than do 'risk-based' approaches. It makes a point of including a wider range of social and political values, rather than those that happen to be embodied in the relatively narrow community of technical specialists.

Precaution involves notions that '*prevention is better than cure*', that '*the polluter should pay*', that options offering simultaneously better economic and environmental performance should always be preferred ('*no regrets*'), that options should be appraised at the level of *production systems* taken as a whole and that attention should be extended to the intrinsic value of *non-human life* in its own right. In effect, precaution is variously taken to mean the adoption of greater *humility* about scientific knowledge, a recognition of the *vulnerability* of the natural environment, the prioritizing of the *rights* of those who stand to be adversely affected by environmental risks. In this way, a 'precautionary approach' introduces an apparently formidable array of additional issues to more conventional 'risk-based' approaches to the appraisal of energy sustainability. This ambiguity, breadth and ambition of precautionary approaches present a number of challenges. In particular, there are concerns in many quarters that a precautionary approach to regulatory appraisal involves the diluting—or even sacrificing—of the clarity, rigour and practical utility of 'risk-based' techniques like cost–benefit analysis and risk assessment.

These issues will be returned to later in this article. First, we examine the extent to which established 'risk-based' approaches to the appraisal of energy sustainability actually succeed in delivering clear, rigorous or practically useful results. The results obtained in a wide literature survey will be reviewed in some detail. A number of serious methodological and theoretical difficulties will be examined. Informed by this discussion, attention will turn to a novel way of thinking about the relationship between 'science' and 'precaution' in the assessment of energy sustainability. Finally, a number of conclusions will be drawn concerning practical ways in which the regulatory appraisal of sustainable energy strategies might at the same time become more scientifically well-founded *and* more precautionary in character.

2   Some Practical Difficulties

Over the past 20 years, there has accumulated an enormous body of literature bearing on the relative environmental impacts—or sustainability—of different electricity supply options. In no other industrial sector is comparative environmental appraisal better established, more extensive or more sophisticated. The particular techniques have changed over time. From the late 1970s to the early 1980s, the dominant analytical approach in the industrialized market economies was provided by comparative risk assessment. From the late 1980s to the mid 1990s, the techniques of environmental cost–benefit analysis (or 'environmental valuation') had begun to complement and, increasingly, substitute the role played by risk assessment. Yet the central objective, many of the methods, and even much of the primary data, have stayed fairly constant. The key task throughout remains the same: to provide a clear picture of the overall magnitudes of the environmental impacts of individual options and a basis for the relative ordering of different options. The purpose of this section will be to review the extent to which these aims have been fulfilled.

Based on a survey of some 32 influential government and industry-sponsored studies conducted in industrialized countries over the past twenty years,[1] Figure 1 displays the results obtained for the aggregate environmental effects of just one option, modern coal power. Adverse effects are here quantified in monetary terms—as 'external costs' (expressed as US cents per kilowatt hour of electricity produced, normalized to 1995 prices). However, essentially the same picture arises in the closely related comparative risk assessment literature, where impacts are expressed in terms of human morbidity and mortality effects. In short, the highest and lowest values derived for the environmental impacts of this energy option vary by more than four orders of magnitude—a factor of more than 50 000.

The variability evident in the literature as a whole far exceeds the uncertainty range expressed in any individual study. This is as true in the comparative risk assessment literature as in the environmental externality literature reviewed in Figure 1. Indeed, many studies present their results as a single numerical value, rather than as a range. The numbers are sometimes expressed with formidable levels of precision. The major Ottinger study for the US Department of Energy,[2] for instance, and a report for the German electricity industry in 1989[3] give some

[1] T. O'Riordan and J. Cameron (eds.) *Interpreting the Precautionary Principle*, Earthscan, London, 1994.

[2] R. L. Ottinger, D. R. Wooley, N. A. Robinson, D. R. Hodas and S. E. Babb, *Environmental Costs of Electricity*, Oceana, New York, 1990.

Figure 1 Variability in the regulatory appraisal of energy options (the case of coal power)

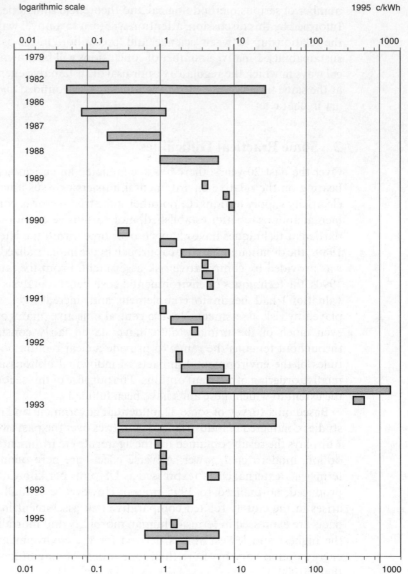

results to three significant figures (one part in one thousand). Similar degrees of precision are adopted in many risk assessment studies.[4–6] The pioneering 1988

[3] A. Voss, R. Friedrich, E. Kallenbach, A. Thoene, H.-H. Rogner and H.-D. Karl, *Externe Kosten der Stromerzeugnung Studie im Auftrag der VDEW*, Frankfurt, 1989.

[4] W. D. Rowe and P. Oterson, *Assessment of Comparative and Non-comparative Factors in Alternate Energy Systems*, Commission of the European Communities, Brussels, 1983.

[5] United Nations Environment Programme, *Comparative Data on the Emissions, Residuals and Health Hazards of Energy Sources*, Environmental Impacts of the Production and Use of Energy, part IV, phase I, UNEP, 1985.

[6] International Atomic Energy Agency, *Senior Expert Symposium on Electricity and the Environment*, key issues papers, IAEA, Vienna, 1991.

Hohmeyer study for the European Commission presents some results to a daunting four significant figures[7,8] (one part in 10 000).

Other studies do acknowledge some degree of uncertainty and variability, by presenting their results as a simple range. In the risk assessment literature, the upper and lower bounds to the ranges of results expressed in individual studies can in some cases extend to several orders of magnitude.[9-12] However, this is far from being a general rule. For the most part, as can be seen in Figure 1, only a rather small degree of uncertainty is acknowledged in the presentation of results. The Hohmeyer study, for instance, yields ranges that are as narrow as factor 10 at most,[13] whilst the Ottinger study mentioned earlier[2] and a study by Pearce for the UK Department of Trade and Industry[14] present no ranges at all in some final results. The 'Externe' environmental cost study for the European Commission adopts a more sophisticated approach to the treatment of data quality, specifying the degree of confidence associated with the values obtained for different disaggregated effects and itself avoiding summing over categories.[15] However, even this study nevertheless presents its results as discrete values rather than as ranges or sensitivities. Where qualifications are buried in the more theoretical passages of such studies, they are all too easily lost in derivative work which treats the results obtained as if they *were* meaningfully additive.[16]

The importance of this treatment of uncertainty is that the apparent precision implied in individual studies masks the enormous variability in the literature as a whole (as shown in Figure 1). When attention turns to a comparison of the externality results obtained for a range of *different* electricity supply options, the ambiguity of the overall picture is further compounded. Based on the same environmental cost literature reviewed in Figure 1,[17] Figure 2 displays the externality values derived in the literature as a whole for eight key generating options. A general pattern emerges under which renewables display generally favourable environmental performance when compared with the fossil fuels, with nuclear power apparently spanning the difference. However, the picture is again

[7] O. Hohmeyer, *Social Costs of Energy Consumption: External Effects of Electricity Generation in the Federal Republic of Germany*, Springer, Berlin, 1988.

[8] O. Hohmeyer, *Latest Results of the International Discussion on the Social Cost of Energy—How Does Wind Compare Today?*, paper presented at the 1990 European Wind Energy Conference, Madrid, October 1990.

[9] C. L. Comar and L. A. Sagan, Health effects of energy production and conversion, *Annu. Rev. Energy*, 1976, **1**.

[10] R. A. D. Ferguson, *Comparative Risks of Electricity Generating Fuel Systems in the UK*, UKAEA, Peregrinus, 1981.

[11] A. F. Fritzsche, The health risks of energy production, *Risk Analysis*, 1989, **9**, no. 4.

[12] J. Holdren, G. Morris, I. Mintzer, Environmental aspects of renewable energy sources, *Annu. Rev. Energy*, 1980, **5**.

[13] O. Hohmeyer, Renewables and the full costs of energy, *Energy Policy*, 1992, **18**, no. 3.

[14] D. Pearce, C. Bann and S. Georgiou, *The Social Cost of the Fuel Cycles*, report to the UK Department of Trade and Industry by the Centre for Social and Economic Research on the Global Environment, HMSO, London, 1992.

[15] European Commission, *Externe: Externalities of Energy*, EUR 16520–16525 EN, Brussels, 1995, vols. 1–6.

[16] Energy for Sustainable Development, ZEW, IARE, IER, Coherence, FhG, *SAFIRE Final Report*, final report for the European Commission DG XII, Corsham, November 1995.

[17] A. Stirling, Limits to the value of environmental costs, *Energy Policy*, 1997, **25**, no. 5.

Figure 2 Ambiguity in the regulatory appraisal of electricity supply options

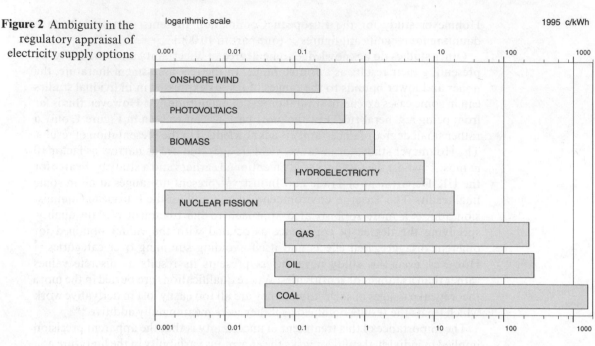

dominated by enormous variability. Individual studies show results at the high end of the overall range for some options but lower in the distributions for others. There is no evident structure to the distributions, nor to the positioning of individual studies. Some results lie at one end of the range for some technologies and at the other end for others. Indeed, since the lowest values obtained for the generally worst-ranking option (coal) are lower than the highest values obtained for the apparently best ranking options (wind), the overall picture would accommodate any conceivable ranking order for these eight options.

Two conclusions may be drawn from this survey of the practical results obtained in quantitative approaches to the appraisal of energy sustainability. The first concerns the notion that these kinds of 'risk-based' approaches deliver a clear basis for policy making. Although individual studies aspire—and even claim—to deliver a 'real',[18,19] 'full'[8,13] or 'true'[2] picture, it can be seen from Figures 1 and 2 that these ambitions conceal an enormous degree of uncertainty, variability and ambiguity. This renders the apparent clarity and precision of the neat quantitative results highly misleading, seriously qualifying the policy utility of the quantitative risk-based approach.

The second conclusion concerns the relative importance of qualitative and quantitative factors in the appraisal of sustainability. The results displayed in Figures 1 and 2 are drawn from a tightly defined body of authoritative studies commissioned by government or industry bodies. Even without including a wider literature reflecting more diverse socio-political perspectives, it is clear that there exists considerable latitude for the exercise of different 'framing assump-

[18] H. Hubbard, The real cost of energy, *Sci. Am.*, 1991, 2644.
[19] G. Hirata and P. Takahashi, *Assessing the Real Cost of Energy*, presentation to conference of the International Association of Energy Economics, July 1991.

tions' in the appraisal of energy sustainability. The real value to policy of formal environmental appraisal methods, therefore, lies not in ostensibly clear simple numerical results, but in the systematic exploration of the way the results obtained (and the relative orderings of options) vary under different—equally legitimate—assumptions.

3 Key Methodological Issues

So what are these crucially important 'framing assumptions' which lie hidden in conventional quantitative environmental appraisal? Essentially, they reflect the intrinsic scope, complexity, ambiguity and subjectivity of the many assumptions that are embedded in the various quantitative methodologies. This is the subject of an enormous critical literature, focusing both on the comparative risk analysis and the later environment cost assessment literature. In order to illustrate the kinds of issues that underlie the variability evident in Figures 1 and 2, the present section will review some of the principle issues that arise.

First, the impacts associated with different generating technologies may differ radically in the *forms* which they take. Some may be more manifest as risks of death, others as injury or disease (*e.g.* offshore wind and wave *versus* biomass). They may differ in the immediacy or latency of their impacts (*e.g.* retrofit rooftop solar arrays *versus* nuclear power). The effects of some options may be concentrated in a few large events, whereas others may spread across a larger number of smaller incidents (*e.g.* nuclear power *versus* coal). Effects of different options may vary in the degree to which they are reversible (*e.g.* nuclear and fossil fuels *versus* wind power). The determination of the relative importance of these different dimensions must inevitably be a highly circumstantial and subjective matter. The aggregated numerical values obtained in 'risk-based' analysis compress these different dimensions onto a single yardstick, adopting implicit assumptions about their relative importance.

Second, the impacts caused by different energy options also differ in terms of their *distribution* across space, through society and is it better that impacts of a given magnitude be geographically concentrated or dispersed (*e.g.* wind *versus* fossil fuels)? This also raises issues concerning the 'fairness' of the distribution of impacts across different groups and the way this correlates (or not) with the distribution of the benefits arising from the operation of the investments concerned. Particularly intractable difficulties emerge in contemplating the distribution of risks through time (*e.g.* nuclear and fossil fuels *versus* renewables), and the balance between burdens which fall on human and non-human life (*e.g.* biomass *versus* gas), workers and the general public (*e.g.* offshore wind *versus* oil) or on communities already affected by other environmental burdens (*e.g.* urban waste to energy *versus* domestic photovoltaics). Where the patterns in the distribution of the risks of different energy options vary along these dimensions, further serious questions must be raised about the value of discrete numerical results, such as those delivered by risk and cost–benefit analysis.

Third, the risks of different electricity supply options also impact differently on the *autonomy* of those affected. Exposure to the effects of some technologies is more voluntary than is the case for others (*e.g.* DIY energy efficiency *versus* centralized coal power). Likewise, different effects vary in their familiarity and the degree to which they are controllable (*e.g.* nuclear *versus* wind). Further serious, complex and pervasive issues are raised in considering the trust that should be placed in the communities and institutions associated with the operation of the different options and the appraisal results which they obtain (*e.g.* nuclear *versus* hydro).

Turning to the characteristics of analysis, rather than the risks themselves, the results obtained in environmental appraisal are, obviously, highly sensitive to the selection of primary quantitative *indicators*. Although final results are (in cost–benefit analysis) expressed as monetary values or (in risk assessment) mortality or morbidity frequencies, these represent conversions and aggregations over a wide variety of basic indices. The different primary metrics employed with each individual risk may vary radically in the degree to which they capture the full character of that individual effect and the fidelity with which they track its dynamics. Some effects are intrinsically much more readily quantifiable than others (*e.g.* particulate emissions *versus* aesthetic landscape effects). This compounds the potential for incoherence between the approaches adopted to different effects, both within individual studies and between different studies.

A further issue is raised in the choice of particular appraisal *methodologies*. In short, different studies tend to emphasize different methods. Some cost–benefit studies obtain their results largely through the pursuit of a 'mitigation cost' approach, based on an assessment of the costs incurred in alleviating environmental damage once committed.[20] Other studies mix results obtained through application of mitigation cost techniques (*e.g.* to certain atmospheric effects) with values obtained by the use of 'hedonic market' and 'contingent valuation' methods (*e.g.* to certain water effects) which assess values, respectively, by examining prevailing property or wage markets or responses to questionnaires.[2] Still other analysts favour the use of 'abatement cost' techniques, which take the costs of controlling pollution at source as a proxy indicator for the social costs of the environmental impacts thereby avoided.[21] A final group of studies is based mainly on a fourth methodology: the 'damage function' technique.[15] This involves the 'bottom-up' assessment of the costs associated with each physical dose–response relationship. There is considerable literature concerning the relative merits and deficiencies of these different techniques.[22–24] The differing characters of these approaches, and the fact that the results obtained for specific effects are found often to vary significantly between methods, are suggestive of important difficulties of resolution, fidelity and coherence in 'risk-based' analysis.

[20] O. Hohmeyer, *Social Costs of Energy Consumption: External Effects of Electricity Generation in the Federal Republic of Germany*, Springer, Berlin, 1988.

[21] Tellus Institute, *Valuation of Environmental Externalities, Sulfur Dioxide and Greenhouse Gases*, Report to the Massachusetts Division of Energy Resources, December 1991.

[22] O. Hohmeyer and R. Ottinger (eds.), *External Environmental Costs of Electric Power Production and Utility Acquisition Analysis and Internalization: Proceedings of a German-American Workshop*, Fraunhofer ISI, Karlsruhe, 1990.

[23] D. Pearce and A. Markandya, *Environmental Policy Benefits: Monetary Evaluation*, OECD, Paris, 1989.

[24] G. L. Peterson, B. L. Driver and R. Gregory (eds.), *Amenity Resource Valuation: Integrating Economics with Other Disciplines*, Venture, Philadelphia, 1988.

A final set of difficulties in the conventional 'risk-based' appraisal of electricity options concerns the fundamental underlying assumptions adopted in the *framing and presentation* of the analysis. In short, assumptions concerning the specific operational circumstances of the different options (*e.g.* with combined heat and power or the use of hydro in water management), their developmental trajectories and the 'system boundaries' set for the purpose of analysis may all have a determining influence on the nature of the results obtained (*e.g.* inclusion of material inputs to renewables or uranium mining). Further issues may be raised in considering the degree to which any individual set of results constitute a 'complete' account of the issues pertaining to any individual decision (*e.g.* with respect to issues like the centralizing of political authority or military and security implications). Crucial questions typically surround the way in which the interpretation of results is complemented by consideration of those factors which are not included in analysis.

To illustrate the potential importance of these 'dimensions of variability' in determining the results of 'risk-based' analysis, examples may be given concerning just two such factors in recent cost–benefit studies of electricity options: the treatment of system boundaries and the completeness in the scope of analysis. The apparently neat numerical values derived in 'risk-based' analysis may conceal the crucial fact that different studies address different stages in the 'fuel cycles' associated with individual options and in the 'life cycles' of associated facilities. Hohmeyer's 1988 study (and its subsequent updates) are essentially restricted to the electricity generation stage (omitting mining or drilling, fuel processing, storage and transport and waste management) and to the operational phase (omitting inputs of energy and materials, and the impacts of the construction and decommissioning processes).[7] The 1990 Ottinger study is almost as restricted in scope, addressing waste management burdens and decommissioning (for some options but not others).[2] The 1992 study for the UK Department of Trade and Industry (DTI) adopts wider system boundaries, including some reference to fuel extraction, processing, transport and storage and waste management, but also omits material and energy inputs and construction and decommissioning impacts.[14] With the exception of the 1995 Externe report,[15] it is notable that the system boundaries set in valuation studies tend to be much narrower than those which have for some time been conventional in the comparative risk assessment of energy options.[25-27] Even the Externe report,[15] however, omits energy and material inputs to construction of fossil and nuclear facilities, while including these for some other options. Crucial underlying assumptions on system boundaries are not conveyed in aggregated numerical results. As a result, 'risk-based' assessments are vulnerable to serious difficulties of interpretation.

A similar picture emerges with respect to the completeness of risk-based

[25] D. J. Ball, L. E. J. Roberts and A. C. D. Simpson, *An Analysis of Electricity Generation Health Risks—a United Kingdom Perspective*, Centre for Risk Assessment, University of East Anglia, Norwich, 1994.

[26] H. Inhaber, *Risk of Energy Production*, AECB-1119/REV-1, Canadian Atomic Energy Control Board, Ottawa, 1978.

[27] N. D. Mortimer, Energy analysis of renewable energy sources, *Energy Policy*, 1991, **17**, no. 4.

methods. Different studies include and exclude different categories of effect. Hohmeyer's 1988 cost–benefit study for the European Commission[7] excludes aesthetic effects, thereby omitting a factor widely regarded as the most serious single environmental impact of wind power. The 1990 Ottinger study[2] does address aesthetic impacts, but omits to account for occupational safety risks, another effect that is sometime argued to be important in assessing wind power.[27] Although relatively comprehensive in scope, the major 1995 Externe study,[15] also conducted for the European Commission, excludes global warming from its final numerical results, despite the fact that this is addressed in both the other earlier studies mentioned. The Externe study also omits to address the possibility of environmental damage due to terrorist attacks or sabotage at nuclear power stations, factors which are elsewhere often viewed as significant.[28] All three studies exclude any attention to the environmental implications of nuclear proliferation, although efforts in this regard are made elsewhere.[29,30] Despite including some of the most thorough and systematic studies in the field, each of these reports, in different ways, may therefore be judged to be seriously incomplete. If risk assessment or cost–benefit results are taken at face value, then this important factor is entirely missed.

When these various issues are taken together, it is an uncomfortable but undeniable fact that the adoption of different but equally reasonable assumptions or conventions on potentially any one of the different dimensions of appraisal may radically affect the results of 'risk-based' analysis of the impacts of different electricity generating options. Yet these crucial assumptions are exogenous to the 'scientific' analysis and essentially contingent and subjective in nature. The more recent cost–benefit approaches seem no more able to avoid this problem than did comparative risk assessment before them. Indeed, based on the examples provided here, it is difficult to escape the conclusion that, by combining additional methodological complexity with apparent presentational simplicity, the more elaborate methods of analysis can make the problem worse rather than better. In short, the notion that orthodox 'scientific' approaches to regulatory appraisal necessarily yield a clear, robust and pragmatic basis for appraisal of energy sustainability seems seriously flawed.

4 Underlying Theoretical Problems

To some, this discussion of variability and inconsistency in the subjective assumptions underlying apparently scientific 'risk-based' approaches to appraisal may seem rather superficial, simply indicating a circumstantial lack of firm methodological disciplines and conventions. Would it not be possible to remove

---

[28] S. Sholly, P. Hofer, A. Gazsó, H. Kromp-Kolb and W. Kromp, *Integrated Risk Assessment for Nuclear Power Plants*, paper presented to the Conference of the European Society for Ecological Economics, Transitions Towards a Sustainable Europe: Ecology, Economy, Policy, Vienna, May 2000, http://www.wu-wien.ac.at/esee2000/.

[29] US Congress Office of Technology Assessment, *Studies of the Environmental Costs of Electricity*, OTA, Washington, DC, September 1994.

[30] M. Shuman and R. Cavanagh, *A Model Conservation and Electric Power Plan for the Pacific Northwest, Appendix 2: Environmental Costs*, Northwest Conservation Act Coalition, Seattle, November 1982.

the difficulties presented for sustainable energy strategies simply by standardizing the methods and procedures of regulatory appraisal? Unfortunately, the theoretical literature shows that the problems are far more deep-seated and intractable than this. Indeed, these difficulties may be taken to reflect the most basic principles in the 'scientific' foundations of probability and rational choice theory: 'incommensurability' (comparing 'apples and oranges') and 'ignorance' ('we don't know what we don't know').

Both issues follow naturally from the discussion in the previous section. Assuming for a moment that the regulatory appraisal literature displayed a level of consistency and completeness far beyond that which is typically achieved, there would still be the question of how the different energy impacts are to be framed and prioritized in sustainability appraisal. For instance, to what extent should analysis be based on well-documented past empirical data relating to possibly outdated options, superseded practices or irrelevant circumstances, or to what extent should it make use of theoretical models of performance based on extrapolations, projections and untested assumptions? How should individual unquantifiable aspects of risk be taken into account? Even where they are fully quantifiable, there is the question of the relative priority that should be attached to the different factors in the aggregation of effects such as toxicity, carcinogenicity, allergenicity, occupational safety, biodiversity or ecological integrity. What relative weight should properly be placed on impacts to different groups, such as workers, children, pregnant and breastfeeding mothers, future generations, disadvantaged communities, foreigners, those who do not benefit from the technology in question or even to animals and plants as beings in their own right?

Even if they were practically feasible, objectives such as completeness or comprehensiveness do not assist in addressing issues of framing and prioritization of this kind. No one set of assumptions or priorities may be claimed to be uniquely rational, complete or comprehensive. It is this which constitutes the problem of incommensurability, a classic and well-explored dilemma in the field of social choice, but one that is frequently forgotten in regulatory appraisal. For it is a fundamental consequence of the axioms of utilitarian rationality that underlie both risk assessment and cost–benefit analysis that neither technique has developed definitive ways to resolve the difficulty of comparing apples and oranges. Even the most optimistic of proponents of rational choice acknowledge that there is no effective way to compare the intensities of preferences displayed by different individuals or groups in society. Indeed, even where social choices are addressed simply in ordinal terms, the economist Kenneth Arrow went a long way towards earning his Nobel Prize for demonstrating formally from first principles within the rational choice paradigm that it is *impossible* definitively to combine relative preference orderings in a plural society.[31]

Put simply, the point is that 'it takes all sorts to make a world'. Different cultural groups, political constituencies or economic interests typically attach different degrees of importance to the different aspects of energy sustainability and look at them differently. Within the bounds defined by the domain of plural social discourse, no one set of values or framings can definitively be ruled more

[31] K. J. Arrow, *Social Choice and Individual Values*, 2nd edn., Wiley, New York, 1963.

'rational' or 'well informed' than many others. Even were there to be complete certainty in the quantification of all the various classes and dimensions of sustainability, it is entirely reasonable that fundamentally different conclusions over environmental risk might be drawn under different—but equally legitimate—perspectives. It is a matter of the 'science' of rational choice itself, then, that there can be no 'analytical fix' for the problems posed by complexity and subjectivity in the appraisal of sustainability. It is ironic that the application of 'scientific' techniques such as risk and cost–benefit analysis should so often neglect such a fundamental result arising from their own underlying 'scientific' first principles.

The second fundamental problem underlying the social appraisal of energy sustainability concerns incomplete information. It is a central feature of 'risk-based' approaches to regulatory assessment that incompleteness in empirical and theoretical knowledge is addressed by applying quantitative probabilistic methods. Indeed, in economics and utility theory, this is the essence of the well-established formal definition of *risk* itself. This is the principal reason for referring to both risk assessment and cost–benefit analysis as 'risk-based' approaches. Here 'risk' is—by definition—a condition under which it is possible both to define a comprehensive set of all possible outcomes *and* to resolve a discrete set of probabilities (or a density function) across this array of possibilities. This is illustrated in the top left-hand corner of Figure 3. It is a domain under which the various techniques of risk assessment are applicable, permitting (in theory) the full characterization and ordering of the different options under appraisal. There are a host of details relating to this picture (such as those hinging on the distinction between 'frequentist' and 'Bayesian' understandings of probability), but none of these alter the fundamental definition of the concept of risk.

The strict sense of the term *uncertainty*, by contrast, applies to a condition under which there is confidence in the completeness of the defined set of outcomes, but where there is acknowledged to exist no valid theoretical or empirical basis confidently to assign probabilities to these outcomes. This is found in the lower left-hand corner of Figure 3. Here, the analytical armoury is less well-developed, with the various sorts of sensitivity and scenario analysis being the best that can usually be managed.[32] Whilst the different options under appraisal may still be broadly characterized, they cannot be ranked even in ordinal terms without some knowledge of the relative likelihoods of the different outcomes.

Both risk and uncertainty, in the strict senses of the terms, require that the different possible outcomes be clearly characterizable or subject to measurement. The discussion in the previous section has already made it clear that this is often not the case: the complexity and scope of the different forms of environmental effect, and the different ways of framing and prioritizing these, can all-too-easily render *ambiguous* the definitive characterization of outcomes (top right corner of Figure 3). Where these problems are combined with the difficulties in applying the concept of probability, we face a condition which is formally defined as *ignorance* (bottom right corner of Figure 3).[33–35] This applies in circumstances

[32] S. Funtowicz and J. Ravetz, *Uncertainty and Quality in Science for Policy*, Kluwer, Amsterdam, 1990.

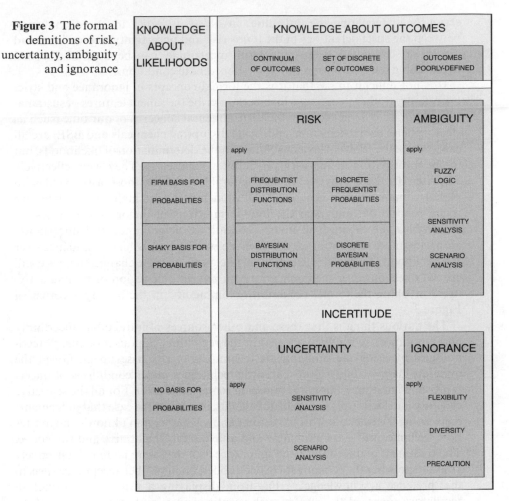

Figure 3 The formal definitions of risk, uncertainty, ambiguity and ignorance

where there not only exists no basis for the assigning of probabilities (as under uncertainty), but where the definition of a complete set of outcomes is also problematic. In short, recognition of the condition of ignorance is an acknowledgement of the possibility of surprise. Under such circumstances, not only is it impossible definitively to rank the different options, but even their full characterization is difficult. Under a state of ignorance (in this strict sense), it is always possible that there are effects (outcomes) which have been entirely excluded from consideration.

Figure 3 provides a schematic illustration of the relationships between these formal definitions for the concepts of risk, uncertainty, ambiguity and ignorance. It is quite normal, even in specialist discussion, for the full breadth and depth of these issues to be conflated in the simple concepts of 'risk' or 'uncertainty', thus

[33] B. Loasby, *Complexity and Ignorance: an Inquiry into Economic Theory and the Practice of Decision Making*, Cambridge University Press, Cambridge, 1976.

[34] M. Smithson, *Ignorance and Uncertainty: Emerging Paradigms*, Springer, New York, 1994.

[35] B. Wynne, *Global Environ. Change*, 1992, **6**, 111–127.

seriously understating the difficulties involved. In order to avoid confusion between the strict definitions of the terms risk and uncertainty as used here, and the looser colloquial usages, the term 'incertitude' can be used in a broad overarching sense to subsume all four subordinate conditions.

It is not difficult to see that it is the formal concepts of ignorance and strict uncertainty (rather than risk) which best describe the salient features of sustainability appraisal. Some of the main environmental concerns of our time (such as stratospheric ozone depletion, endocrine disrupting chemicals and BSE) are all cases where the problem lay not so much in the determination of likelihoods, but in the anticipation of the very possibilities themselves. They were effectively surprises. In the energy sector, imponderables such as those associated with global climate change, geological diffusion models for high-level radioactive waste repositories and even the long-term effects of major dependencies on renewables like biomass all invoke conditions of ignorance and uncertainty alongside more straightforward risk. Even where there is high confidence over the likelihood of a broad phenomenon, like global climate change, there are still crucial questions over the implications for any specific region or human activity,[36] invoking the formal condition of 'ambiguity' in the top-right corner of Figure 3.

The curious thing is that these and other sources of intractable uncertainty and ignorance are routinely treated in the regulatory appraisal of energy technologies by using the probabilistic techniques of risk assessment. Given the manifest inapplicability of probabilistic techniques under conditions of uncertainty and ignorance, this is a serious and remarkable error. For all the seductive elegance and facility of probabilistic calculus, it remains the case that judgements concerning the extent to which 'we don't know what we don't know'—no matter how well informed—are ultimately and unavoidably qualitative and subjective. The treatment of uncertainty and ignorance as if they were mere risk effectively amounts to what the economist Hayek dubbed (in his Nobel acceptance speech) the 'pretence at knowledge'.[37] Far from displaying a respect for science in regulatory appraisal, the effect of such scientistic oversimplification is actually to ignore and undermine scientific principles.

Both with respect to 'incommensurability' and 'ignorance', then, it is clear that the aspiration to definitive prescriptive conclusions through 'risk-based' approaches to regulatory appraisal is not only hitherto unachieved in regulatory appraisal, is it just unfeasible in practice. In a plural society, the asserting of an unequivocally 'sound scientific' basis for particular sustainable energy strategies is a fundamental contradiction in terms!

5 Science and Precaution in Energy Sustainability

It is with increasing realization of these practical and theoretical limitations to the value of 'risk-based' techniques in the appraisal of sustainability that interest

---

[36] Intergovernmental Panel on Climate Change, 2, *Second Assessment Report*, Oxford University Press, Oxford, 1995.

[37] F. von Hayek, *New Studies in Philosophy, Politics, Economics and the History of Ideas*, Chicago University Press, Chicago, 1978.

is growing in complementary and alternative approaches, such as the pre-cautionary approach discussed in the Introduction. In particular, the Pre-cautionary Principle has become embodied in an increasing number of national and international statutory instruments over recent years.[1] As Article 15 of the 1992 Rio Declaration, the Precautionary Principle is established as a key el-ement in mainstream notions of sustainability. This classic formulation holds that: '*Where there are threats of serious or irreversible damage, lack of full scientific certainty shall not be used as a reason for postponing cost-effective measures to prevent environmental degradation*'.

Although more narrowly framed than the wider array of issues informing the precautionary approach and reviewed at the beginning of this article, it is clear from the outset that the Precautionary Principle is not without its own intrinsic ambiguities. For instance, questions may be raised over the precise interpreta-tion of concepts such as 'threat', 'irreversibility', 'full scientific certainty' and 'cost-effectiveness' in the typical formulation given here. What do these terms actually mean for concrete choices between energy strategies? To some, the resulting potential for confusion and arbitrariness is seen as a sign of a ten-sion—or even contradiction—between the characteristics of 'precautionary' pol-icy and the discipline and rigour thought to be associated with 'scientific' approaches to regulatory appraisal. Indeed, the view has been expressed in some quarters that a 'precautionary approach' is so lacking in its theoretical frame-work, and so ambiguous in its practical implications, that it is of little or no policy value. At the extreme, it is feared, implementation of 'precaution' would militate against the adoption of any new technology at all and so represents a retreat from deeply held convictions over the positive potential of technological progress. Nowhere is this more pertinent than in the energy sector, where pressing challenges such as climate change provide a strong imperative for radical innovation and profound technological transitions.

Of course, the central message of the preceding sections of this article is that these concerns do not address the fact that orthodox 'scientific' approaches like comparative risk assessment and cost–benefit analysis are actually themselves inconsistent with scientific principles of rational choice, present serious method-ological ambiguities and yield wide-ranging and ambiguous results. Seen in this light, precaution presents no difficulties that are not already intrinsic features of appraisal. Indeed, by providing for greater acknowledgement of these intrinsic contingencies and uncertainties in appraisal—it can be argued that precaution does go usefully beyond conventional risk-based approaches. The Precautionary Principle itself, for instance, provides explicit guidance that the inevitable uncer-tainties in appraisal be highlighted rather than marginalized or denied. Further-more, it provides for a specific response, in that greater benefit of the doubt be granted to the environment and to public health than to the activities which may be held to threaten these things. A host of different practical regulatory instru-ments and measures are variously proposed in different contexts as following from this.[38,39]

[38] A. Stirling, *Precautionary and Science-Based Approaches to Risk Assessment and Environmental Appraisal*, Institute for Prospective Technology Studies, Seville, 1999, available at: ftp://ftp.jrc.es/pub/EURdoc/eur19056enII.pdf.

Beyond this, however, the broader elements of the precautionary approach reviewed at the beginning of this article add a further dimension to the appraisal of sustainable energy strategies. Rather than being seen in terms of an ostensibly simple 'decision rule' embodied in a formulaic legal principle (much like the methodological conventions of risk assessment), precaution might instead be seen as a feature of the *process* in which decisions take place. Drawing on a wide literature,[40] these processual elements of a 'precautionary approach' might be summarized as follows.

First, that the scope of appraisal should extend to all those types of effect that are held to be relevant, including qualitative as well as quantitative issues and indirect as well as direct effects. It should not be restricted to highlighting those impacts that are most readily quantified by a particular method, or to treating certain effects as a proxy for others. The presentation of results should fully acknowledge the uncertainties and retain a high degree of humility about the potential for ignorance and surprise.

Second, that attention should extend to the benefits and justifications associated with different options as well as the risks and costs. This enables the 'pros' to be considered alongside the 'cons'. Sustainability intrinsically involves a balance between the pursuit of social and environmental aims, whilst reducing other adverse effects. Risk assessment focuses on adverse effects on a case-by-case basis. Cost–benefit analysis requires that all effects be assessed in terms of monetary values. By providing for consideration of a broader array of options and their respective pros and cons, a precautionary approach may actually serve to facilitate more innovative choices.

Third, that appraisal should not implicitly embody the particular value judgements and framing assumptions of a narrow group of specialists. It should accommodate a diverse array of different points of view (including, importantly, those of potential 'victims') and anticipate a wide range of possibilities in the face of uncertainty and ignorance. This will allow for the systematic testing of sensitivities to different 'framing assumptions', allowing for greater transparency in appraisal and greater accountability on the part of subsequent processes of decision making.

Once these specific elements of a precautionary approach have been identified, it is interesting to return to the question raised at the outset: the extent to which 'scientific' and 'precautionary' approaches in the appraisal of sustainability are actually necessarily in tension. Of course, just as is the case with 'precaution', the meaning of the term 'scientific' is hotly contested and open to all kinds of rhetorical and expedient interpretations. However, for present purposes it is possible relatively uncontroversially to identify a series of distinguishing characteristics of what constitutes 'scientific' appraisal. These may not always be realized in practice, but they present ideal aims that should be aspired to. They are discussed in more detail by the present author elsewhere.[40] In short, a

---

[39] C. Raffensberger and J. Tickner, *Protecting Public Health and the Environment: Implementing the Precautionary Principle*, Island Press, Washington, 1999.

[40] A. Stirling, *On Science and Precaution in the Management of Technological Risk*, Institute for Prospective Technology Studies, Seville, 1999, available at: ftp://ftp.jrc.es/pub/EURdoc/eur19056en.pdf.

Figure 4 A model of the relationships between the concepts of risk, science and precaution

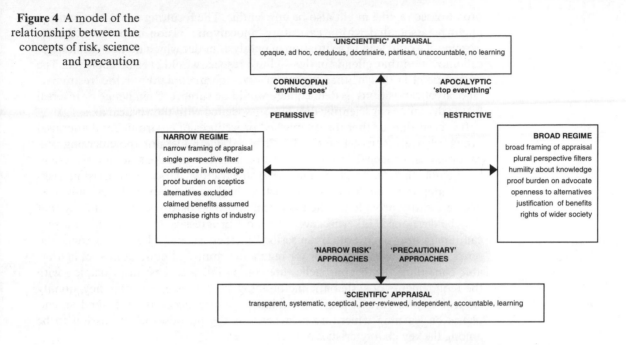

scientific approach to the assessment of the relative sustainability of different energy options should—ideally and at minimum—be *transparent* in its argumentation, *systematic* in its analytical methods, *sceptical* in its treatment of knowledge claims, subject to *peer review*, *independent* from special interests, professionally and democratically *accountable* and continually open to *learning* in the face of new knowledge.

What is interesting about the preceding characterizations is that they reveal an inherently consistent—and in many respects complementary—relationship between 'precaution' and 'science' in regulatory appraisal. Rather than being in tension with the elements of a scientific approach, the key features of precaution serve rather to complement them, in particular by extending the 'breadth of framing' of regulatory appraisal. Figure 4 distinguishes between different approaches to regulatory appraisal based on the degree to which each embodies the respective characteristics of 'scientific appraisal' and 'breadth of framing' identified here. Of course, both the 'broad'/'narrow' and the 'scientific'/'unscientific' distinctions drawn here are highly stylized and simplified. However, the broad picture revealed in Figure 4 is at least richer and more realistic than any one-dimensional dichotomy between 'science' and 'precaution'. Taken together, the combination of these two dichotomies generates the fourfold array of idealized permutations displayed in Figure 4. The adoption of a 'narrow' regime without reference to scientific understandings or disciplines in appraisal might be described as a *permissive* position. Taken to an extreme, this would amount to an entirely uncritical 'anything goes' approach to the regulation of technology of the kind associated with caricature 'cornucopian' visions of progress. Similarly, a broad-based regime might also be unscientific. The resulting *restrictive* position might be associated with a caricature 'apocalyptic' vision of progress. In the extreme, it would lead to a situation of paralysis under which no new technological innovation that offends in the slightest respect would ever be deployed. The crucial point is that neither the 'permissive' (cornucopian) nor the 'restrictive' (apocalyptic) positions as defined here would be subject to challenge or reversal by the disciplines of scientific discourse associated with the vertical axis.

It is clear that neither the established procedures of environmental appraisal (using relatively narrowly framed 'risk-based' methods) nor the emerging precautionary approach (based on broader perspectives and considerations) actually resemble these stylized 'permissive' or 'restrictive' caricatures. Existing risk-based appraisal includes a host of effective checks and balances. It certainly does not necessarily provide for the uncritical approval of any new technology that may be developed. Likewise, even the most progressive formulations of a precautionary approach are circumscribed in their scope, admit an incremental series of instruments and allow for regulatory approval under a host of favourable conditions. Both approaches are compatible—at least in principle—with the requirements of systematic methodology, scepticism, transparency, quality control by peer-review, professional independence, accountability, and an emphasis on learning which are held here for the purposes of discussion to be among the key characteristics of science.

It is at this point that it is useful to return to the earlier discussion in this chapter of the profound importance of the condition of ignorance, and incommensurability in regulatory appraisal. It was shown there that questions over the scope of appraisal, the plurality of different value positions and framing assumptions, the diversity of different anticipated possibilities and the degree of confidence placed in the available knowledge are all matters that are central to the 'scientific' status of the appraisal process. It flows directly from the theoretical foundations of risk assessment, cost–benefit analysis (and, indeed, all rational choice approaches to decision making on risk) that probabilistic approaches are inapplicable under strict uncertainty and ignorance. It also follows equally directly from these fundamental theoretical principles that different priorities, framing assumptions and value systems cannot be definitively aggregated across divergent social perspectives. For both these reasons, it is clear that there can be no analytical fix in assessing the sustainability of a series of different technology or policy options. All that can be done to respect principles of scientific rigour in appraisal is to ensure that the process is as broadly framed as possible in terms of the value systems and framing assumptions that are included and the options and possibilities that are addressed. Seen in this way, then, key elements of the 'breadth' of the regulatory regime themselves constitute issues of 'sound science' in regulatory appraisal. It is for these reasons that—by virtue of its broader basis—the precautionary approach is represented in Figure 4 as being somewhat *more* scientific than the traditional 'narrow' approach involving techniques like risk assessment and cost–benefit analysis.

6 General Implications for the Development of Sustainable Energy Strategies

Although the comparative analysis of the relationships between 'science' and 'precaution' in the appraisal of sustainability may be somewhat novel, the general message of this article is not new. Over recent years, the idea that different forms of environmental and health effect might fruitfully be compared in objectively determinate quantitative terms has fallen under serious doubt. Bodies such as the European Commission,[41] the US National Research Council,[42] the British Royal Commission on Environmental Pollution[43] and even the UK Treasury[44] have come to acknowledge the intrinsically subjective (and thence political) character of regulatory appraisal. Whilst specialists may often reasonably claim greater authority with respect to the assessment of the likely probabilities or physical magnitudes of precisely specified *individual* effects, it is increasingly recognized that expert judgements are as essentially subjective as any other when it comes to the relative prioritization of *different* effects. This is particularly the case where decisions are subject to the condition of ignorance. A detailed set of general recommendations for jointly implementing the disciplines of 'science' and 'precaution' in the appraisal of sustainability are discussed elsewhere.[40] The present article will close by highlighting a few of these general principles in the specific context of energy technologies.

The critique of 'risk-based' appraisal surveyed here has the effect of qualifying the 'objective' status and utility of ostensibly precise quantitative appraisal techniques. However, it should not be concluded that this newly emerging consensus requires the complete abandonment of the discipline and clarity of quantitative methods. The implication is simply that they be treated as 'tools' rather than as 'fixes'. Once we are prepared to relinquish the aspiration to definitive prescriptive results, the key features of a more realistic approach to the appraisal of sustainability become quite clear. For instance, simply by treating technological risk as a vector (rather than a scalar) quantity, straightforward multi-criteria techniques permit a more systematic approach to the multi-dimensional character of environmental effects. Likewise, numerous tools exist for the substitution of single values expressed to several significant figures with systematic sensitivity analysis. Finally, it may be that the problem of divergent assumptions, values and uncertainty might also be addressed by the adoption of a rigorous approach to diversification: focusing on portfolios as a whole, rather than on the 'first-past-the-post' identification of the 'best' individual options.[45,46]

[41] European Commission, *Communication on the Precautionary Principle*, COM(2000)1, Brussels, February 2000.

[42] P. Stern and H. Fineberg, *Understanding Risk: Informing Decisions in a Democratic Society*, US National Research Council Committee on Risk Characterization, National Academy Press, Washington, DC, 1996.

[43] Royal Commission on Environmental Pollution, *Setting Environmental Standards*, HMSO, London, 1998.

[44] Inter-Departmental Liaison Group on Risk Assessment, *The Setting of Safety Standards: a Report by an Interdepartmental Group and External Advisers*, HM Treasury, London, June 1996.

[45] A. Stirling, Diversity and ignorance in electricity supply investments, *Energy Policy*, 1994, **22**, no. 3.

Either way, it is clear that an essential but hitherto neglected input to regulatory appraisal is the transparent inclusion of divergent public perspectives and value judgements. In this light, the need for active public participation in the analysis underlying risk regulation is not simply a question of democratic accountability and political legitimacy. It is a fundamental matter of analytical rigour.[47]

In response to this emerging new climate, a large array of new procedures are under development in many countries for enabling the efficient inclusion of divergent social interests and values right at the outset in regulatory appraisal. These include consensus conferences, citizen's advisory panels, citizen's juries and focus groups.[48-50] Although valuable experiments have been conducted in many areas, such techniques have for the most part yet to be seriously pursued on a large scale as a means to inform real policy decisions. This is just as true in the energy sector as in the wider economy, and applies as much to the allocation of research and development funds, the development of new technologies, their implementation in energy infrastructures and their siting in the landscape.

So what might the social appraisal of sustainable energy options actually look like, were it to be based on comprehensive and systematic sensitivity analysis under a systematic open framework addressing diverse portfolios rather than individual options and including different viewpoints? A pilot of this type of approach in the hotly contested area of genetic modification is reported elsewhere.[51] Drawing on an earlier schematic study in the energy sector,[52] Figure 5 displays as a set of pie charts the implications for the UK generating mix of adopting a range of perspectives concerning the framing and relative importance of different appraisal criteria. This stylized and purely illustrative exercise models the appraisal of three groups of UK generating options (nuclear, fossil fuels and renewables) under three major classes of environmental risk (land use, air pollution and 'nuclear issues'). In addition, account is taken of the economic performance of the different options under prevailing market conditions, and of the possibility of deliberately retaining some diversity in the generating mix as a whole. Based on a systematic set of permutations, the 81 different pie charts each represent an electricity supply mix which would be 'optimal' for the UK under a particular set of framings and weightings on the various appraisal criteria, taking account of cost, environmental performance and operational factors.

To the extent that it employs real technical performance data under each criterion and to the extent that the overall range of weighting schemes might be

[46] A. Stirling, Optimising UK electricity portfolio diversity, in G. MacKerron and P. Pearson (eds.), *The UK Energy Experience: a Model or a Warning?*, Imperial College Press, London, 1996.

[47] A. Stirling, Risk at a turning point, *J. Risk Res.*, 1998, **1**, no. 2.

[48] S. Joss and J. Durant, *Public Participation in Science: the Role of Consensus Conferences in Europe*, Science Museum, London, 1995.

[49] O. Renn, T. Webler, H. Rakel, P. Dienel and B. Johnson, *Policy Sci.*, 1993, **26**, 189.

[50] O. Renn, *Fairness and Competence in Citizen Participation*, Kluwer, Amsterdam, 1996.

[51] A. Stirling and S. Mayer, *Rethinking Risk: a Pilot Multi-criteria Mapping of a Genetically Modified Crop in Agricultural Systems in the UK*, report for the UK Roundtable on Genetic Modification, SPRU, University of Sussex, 1999.

[52] A. Stirling, Multicriteria mapping: mitigating the problems of environmental valuation?, in J. Foster (ed.), *Valuing Nature: Economics, Ethics and Environment*, Routledge, London, 1997.

Figure 5 An illustrative
multi-criteria 'sensitivity
map', based on a
hypothetical exercise

KEY:
■ fossil fuels   ■ nuclear power   □ renewable energy

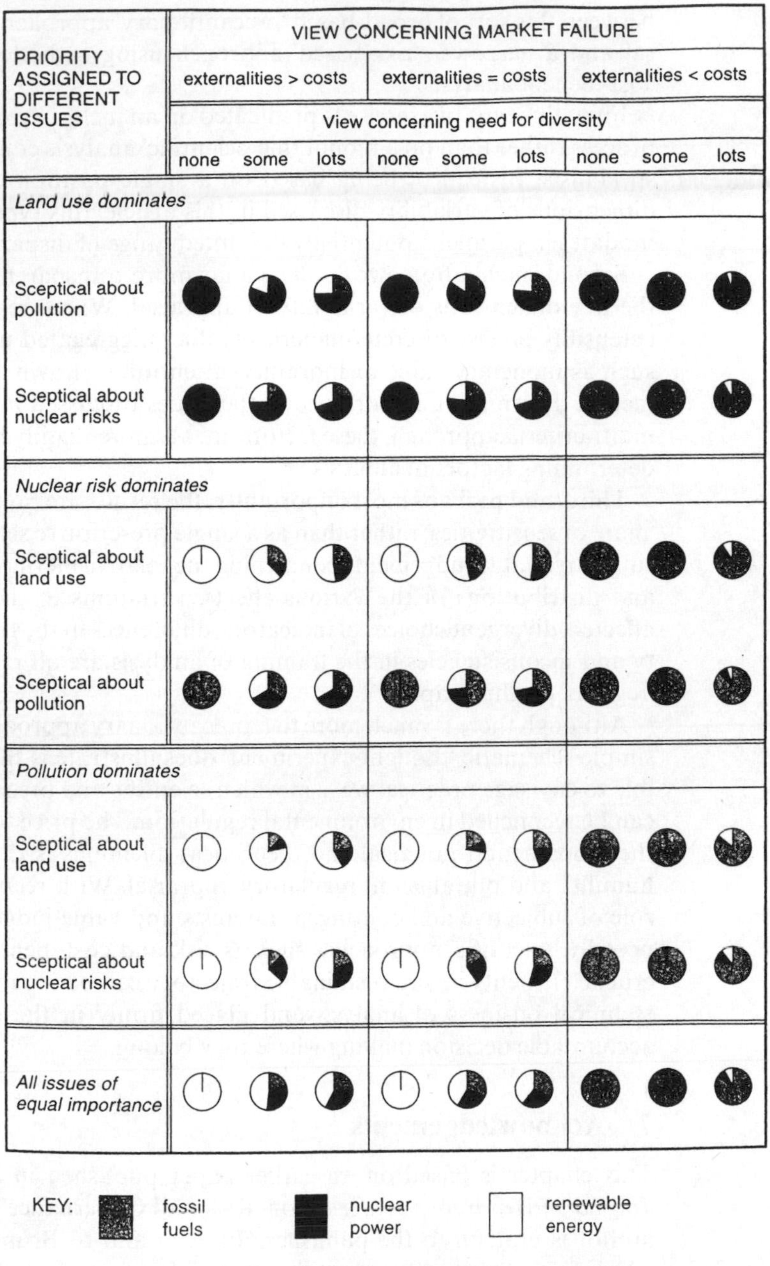

held to accommodate a large portion of the present energy debate, this hypothetical exercise might be viewed as a very rough first-order approximation of what the results of a real empirically based appraisal might look like, were it to be undertaken as part of a formal regulatory appraisal process, or were it to be substituted for one of the many major officially sponsored regulatory appraisal studies reviewed in this article. Although just a schematic reflection of a hypothetical exercise, Figure 5 does serve to illustrate a number of key differences between this sort of broad-based 'precautionary' approach to regulatory appraisal and a narrower 'risk-based' approach using orthodox risk assessment or cost–benefit analysis.

First, this type of exercise is predicated on an inclusive participatory appraisal process rather than on a monolithic 'scientific' analysis conducted exclusively by specialists. Instead of being based on a single position concerning the many dimensions of variability discussed in this article, this type of appraisal accommodates in parallel a potentially unlimited range of disparate positions.

Second, such a framework offers a far more transparent way of dealing with the key dimensions of variability in appraisal. Where results are presented as ostensibly precise discrete numerical values, aggregated under familiar metrics such as monetary value or mortality, attention is drawn away from the fundamental determining importance of the issues discussed in this article. Under a multi-criteria approach, these factors are all more readily highlighted as the key determining factors in analysis.

Third, and perhaps most importantly, the results are presented as a systematic 'map' of sensitivities, rather than as a single prescriptive set of values. Essentially subjective value judgements concerning the relative merits of the disparate forms and distributions of the various effects, variations in the autonomy of those affected, divergent choices of indicators, differences in the treatment of uncertainty and inconsistencies in the framing of analysis are all represented as different 'regions' on this map.

Although there is much more to a precautionary approach than this, what this simple schematic 'thought experiment' does illustrate is that it is perfectly possible to envisage practical ways in which scientific and precautionary imperatives can be reconciled in environmental regulation. The price to be paid for escaping the fundamental practical and theoretical dilemmas is the adoption of greater humility and pluralism in regulatory appraisal. With recognition of the central role of subjective and contingent interests and value judgements—even in supposedly 'scientific' approaches such as risk and cost–benefit assessment—these crucial elements in any sustainable trajectory can be separated from the narrow technical business of analysis and placed firmly in the domain of politically accountable decision making where they belong.

7   Acknowledgements

This chapter is based on an earlier paper published in a special issue of the *Journal of Hazardous Materials* on 'Risk and Governance', (2001, **86**, 5–75). The author is grateful to the publishers Elsevier and to Bruna de Marchi as guest editor for permission to adapt this material.

Subject Index